# Creative Confidence and Music Production

*Creative Confidence and Music Production: Overcome Your Insecurities* is a practical guide for music producers to tackle self-doubt and navigate the relationship between confidence and creativity, by offering tools to overcome the most common creative blocks.

The book begins by discussing the interrelationships between confidence, creativity, and working with technology. This is followed by chapters featuring expert tips and practical exercises to help readers overcome challenges such as struggling with confidence in their production process, and navigating the music industry as an artist. Also included are sections that focus on creative music production workflows, providing practical tips on how to find creative direction and how to work through creative blocks, before finishing with real stories from a range of diverse music professionals about their own workflows, what inspires them, and how they overcome doubt, uncertainty, and lack of inspiration.

Although mainly aimed at music producers, *Creative Confidence and Music Production* has plenty of insights for anyone in the music industry, and can help beginners, music production students, and seasoned professionals alike, to face their fears and foster confidence in their practice.

**Liina Turtonen** (LNA) is an artist, music producer, and sound designer renowned for her chill house music and innovative interactive sound design. An Ableton Certified Trainer and former music production lecturer, she is best known for her popular YouTube channel, "LNA Does Audio Stuff".

# Creative Confidence and Music Production

Overcome Your Insecurities

**LIINA TURTONEN**

LONDON AND NEW YORK

Designed cover image: Emma Holdway

First published 2025
by Routledge
4 Park Square, Milton Park, Abingdon, Oxon OX14 4RN

and by Routledge
605 Third Avenue, New York, NY 10158

*Routledge is an imprint of the Taylor & Francis Group, an informa business*

© 2025 Liina Turtonen

The right of Liina Turtonen to be identified as author of this work has been asserted in accordance with sections 77 and 78 of the Copyright, Designs and Patents Act 1988.

All rights reserved. No part of this book may be reprinted or reproduced or utilised in any form or by any electronic, mechanical, or other means, now known or hereafter invented, including photocopying and recording, or in any information storage or retrieval system, without permission in writing from the publishers.

*Trademark notice*: Product or corporate names may be trademarks or registered trademarks, and are used only for identification and explanation without intent to infringe.

*British Library Cataloguing-in-Publication Data*
A catalogue record for this book is available from the British Library

*Library of Congress Cataloging-in-Publication Data*
Names: Turtonen, Liina, author.
Title: Creative confidence and music production : overcome your insecurities / Liina Turtonen.
Description: Abingdon, Oxon ; New York, NY : Routledge, 2025. | Includes bibliographical references and index.
Identifiers: LCCN 2024046549 (print) | LCCN 2024046550 (ebook) | ISBN 9781032047416 (hardback) | ISBN 9781032047409 (paperback) | ISBN 9781003194484 (ebook)
Subjects: LCSH: Music—Production and direction—Vocational guidance. | Music—Vocational guidance. | Musicians—Interviews.
Classification: LCC ML3795 .T93 2025 (print) | LCC ML3795 (ebook) | DDC 780.23—dc23/eng/20241003
LC record available at https://lccn.loc.gov/2024046549
LC ebook record available at https://lccn.loc.gov/2024046550

ISBN: 978-1-032-04741-6 (hbk)
ISBN: 978-1-032-04740-9 (pbk)
ISBN: 978-1-003-19448-4 (ebk)

DOI: 10.4324/9781003194484

Typeset in Dante and Avenir
by codeMantra

Access the Support Material: www.creative-confidence.com

# Contents

|   | Introduction | 1 |
|---|---|---|
| 1 | Creativity, Confidence, and Music Production | 5 |
| 2 | Creativity, Confidence, and Technology | 23 |
| 3 | Insecurity Corner: Introduction | 43 |
| 3.1 | I Do Not Have a Techy Brain | 47 |
| 3.2 | Audio and Music Production Is Difficult | 53 |
| 3.3 | I Am Too Different to Fit In | 59 |
| 3.4 | My Music Does Not Fit into Any Existing Genre | 65 |
| 3.5 | I Do Not Know Music Theory so I Cannot Learn Music Production | 73 |
| 3.6 | I Do Not Have Enough Money and the Right Equipment | 78 |
| 3.7 | There Is a Right and Wrong Way to Make Music | 85 |
| 3.8 | I Do Not Have Time to Learn and Create | 91 |
| 3.9 | I Feel the Pressure of Being Perfect | 97 |

| | | |
|---|---|---|
| 3.10 | I Feel Jealousy and Envy | 105 |
| 3.11 | I Cannot Stop Asking for Validation | 111 |
| 3.12 | I Fear Failure | 121 |
| 4 | Workflow Theory | 127 |
| 4.1 | Get to Your Goals with Workflows | 135 |
| 4.2 | Workflow Examples | 147 |
| 5 | Industry Professionals Share | 154 |
| | *Acknowledgements* | 173 |
| | *Glossary* | 174 |
| | *Index* | 179 |

# Introduction

I am neurodiverse, an immigrant, a woman, and a school failure, and I am writing this book about creative confidence and music production. So, I am happy to let you know, my dear reader, that this book is not written by a confident person.

Everything about writing this book makes me feel insecure. Maybe I cannot write well enough English for academic standards, or perhaps people will not read it. I might just be an imposter giving myself too much credit. But what is the alternative? Give into my insecurities, create a reality in my head where someone like me could never be an author and settle for that without trying, echoing with the disappointment of things I never dared to achieve. Or, I can try.

As people, we all have desires for things like feeling secure, fitting in, being respected, and getting acknowledged.[1] These needs can inspire us to chase our aspirations or hold us back. Therefore, if we experience an imbalance in any of them, we might be required to confront and examine our insecurities to achieve the necessary balance, which can be daunting. It is important to normalise these emotions and develop the ability to work on our fears and any defences we might give to ourselves.

Having good confidence does not mean that we have everything figured out. It means we are willing to face whatever comes our way. To overcome our insecurities, we need to love ourselves first and look within to identify what makes us feel insecure. For instance, I spent years confused about my identity as a musician, artist, and engineer, but I realised that the answers I was looking for would not come from the industry. Instead, I had to find them within myself. So I believe that by talking openly about our emotions, we can develop a better understanding of how to find our place in the music industry and continue to grow creatively to reach our full potential.

With *full potential*, I do not mean we always need to be better or constantly achieve something. For me, it means stopping self-limiting and self-destruction and understanding what you truly want in life. For all of us, full potential could look different; it could be about slowing down, finding balance in all that we do, or it could be about gaining the courage to leap towards dreams that otherwise would not be lived.

The topic of this book intrigues me, as for a long time I have wanted to gain confidence in my artistry. With this book I analyse creativity and its impact on me as an artist in the modern music industry. So, I have committed to sharing some of my own personal experiences throughout the chapters, because how could I possibly write a book about such a sensitive subject without first being brutally honest about my own insecurities. It is a challenging position to put myself in, as we all have our own biases. I recognise the privilege I have been afforded in certain aspects of my life, and I try approach the topics in this book acknowledging that everyone's experiences are unique. These are just an insight into my experiences, which I hope can be conversation starters and something for readers to reflect on. The discussion continues beyond this book via a YouTube channel and a podcast, to share more perspectives from the creative community. The links to both of these can be found at the end of this chapter.

It is possible that all of this talk about *loving yourself, healing*, and maybe even some of the exercises suggested, might come across as "cringe" to some of you. You might feel that discussions about emotions are uncomfortable or that none of the practical tips feel realistic for you to accomplish at this moment in time, and it is okay to feel that way. I do recognise that for some, looking into what causes our insecurities is not something you wish to do at this time. Maybe some of you are tired of going over traumatic memories, and some want to move on to live brighter days. Wherever you are in your journey, I hope you find something from this book that will give you inspiration, solace, or the help you need to get wherever you wish to be. That is why this book is designed in a way that anyone can take from it what they need, at any stage of their life.

Chapters 1 and 2 discuss confidence, creativity, and their relationship to technology. Chapter 3, called *Insecurity Corner*, was inspired by sentences I have heard from colleagues and students in my line of work over the years. For example, "I just do not have a techy brain" or "I never finish any songs because I am a perfectionist". I wanted to discuss the most common insecurities of the music and audio industries in one chapter, as it is likely that every person reading this book can somehow relate to them.

Chapter 4, about workflows and plans, is for anyone looking for faster and more practical techniques to overcome creative blocks while making music. As I have mentioned above, some readers may feel it is not for them to dive deeper into their emotions or insecurities. With this in mind, this chapter is there to give guidance and direction on a more practical level, where emotional deep-dives are not needed.

In Chapter 5, you can read, relate, and get inspired by ten fantastic music producers, engineers, musicians, artists, and creatives. A lot of our insecurities come from levels of imagination. We often put others on a pedestal and believe that they do not feel the same insecurities as us. This is why I thought it was essential to interview people from the music and audio industries to share their stories. Maybe we are more alike than we initially imagined. These are people from all levels of their careers and different backgrounds. They share what obstacles they have needed to overcome in their lives and how they overcame them, in the hope that they can inspire and help others with their journeys.

The topics discussed in this book can be relatable for people from all aspects of the music industry and creative industries. This book is aimed at music producers and people in audio specifically, due to the nature of the industry and how quiet it is about the insecurities we all experience.

As technical people, we love to have rules for making something function, so the process can get extremely frustrating when our creativity does not work the way we want. So I understand why video tutorials with topics, such as "How to make a hit song" or "5 ways to overcome your creative blocks", are remarkably popular. We seek instructions for our minds and we wish there was a shortcut for our creativity. Would it not be wonderful to know the recipe for a hit song that would sort out your career for the rest of your life?

Although we cannot program our brains and hearts like a piece of software, we can find clarity in our creativity through self-discovery. Understanding what makes you tick and what prevents your full creative potential is maybe what you need, so you will no longer be lost in the depths of your creative mind or seek shortcuts.

Therefore, in this book, we leave the shortcuts, devices, and gadgets backstage and focus on ourselves as the primary tool for our creativity. I am not in any position to tell you what is right for you, or how you feel and experience anything. But I can share the honest reflection of my own insecurities with music production and my career, and I invite others in the industry to do the same. The more we share, the more we can all have the courage to navigate our journeys through emotions such as validation, success, fear, and failure. The process for creative self-discovery can only free more space for our artistry and grow our confidence in the process.

## 4  Creative Confidence and Music Production

> ### DEFINITION OF A MUSIC PRODUCER
>
> The modern definition of music production is broad and complicated. Some people practise it from a purely technical perspective or as the creative directors of musicians' work. Then there is the most current definition where we have entered a world where producers are not purely engineers, but artists using technology as their creative tool. Nevertheless, something that combines all of them is the need for creative confidence, which is not purely about the courage within our own aesthetic choices, but equally about balancing between expectations of the music industry, society, and the technological world of audio.
>
> The scattered definition of the concept of *Music Producer* has created a job title that combines the pressures of two industries: the music industry and the tech industry. So no wonder why the most asked questions, for an educator like me, are about achieving success while being authentic, creative, and excelling in all of the following: music theory, audio physics, technological equipment, and the music business.
>
> In this book, my definition of a music producer is a musician or artist who uses electronic software, instruments, and hardware to produce music for others or themselves. They might be a bedroom producer or one working in a big studio. Music producers might be performing their songs using a computer or electronic instruments, or performing the tracks as a DJ. Overall, the music producers in this book are artists creating music using technology as their primary tool.

The conversation continues on my podcast and YouTube channel called "I Was Just Thinking". Please find more info about them and extra material from: www.creative-confidence.com.

If you wish to follow me as an artist, listen to my music or learn from my YouTube tutorials head to: www.lnamusic.com.

Or find me from social media with the tag @LNADoesAudioStuff

### Note

1  Saul Mcleod. 2020. "Maslow's Hierarchy of Needs". *Simply Psychology*, n.p.

# Creativity, Confidence, and Music Production

## Keywords

- creativity
- confidence
- insecurity
- flow
- brain waves
- workflow
- divergent thinking
- convergent thinking
- meditation

## Music Producers and Creative Confidence

It takes confidence to give yourself time to sit down and open up your computer. It takes determination to believe that you can learn something unexpected, believe in your passions, and keep improving your skills. It takes nerve to decide to pursue something that seems impossible and keep getting up when you fall. And in the midst of this, it takes courage to believe in your ideas when others do not, and create, innovate, and stay true to your vision. It takes confidence to be a music producer.

What surprised me the most about becoming a music production educator was not how many students showed up to the class stoned or how obsessed everyone was about learning sidechaining, it was how the most common questions were about seeking creative confidence. A student struggling to focus on a synthesis lesson might have had previous bad experiences in science lessons, so learning big concepts like that can seem terrifying. Similarly, the amount of choice in samples, instruments, and tools can overwhelm another student, and they cannot get any music down despite concentrating and studying every topic carefully. These are just two examples in a sea of ways that our unique insecurities manifest in our learning and creative practice.

Given the volume of conversation online and in the classroom around these issues, I see a need to find solutions to our blocked creativity and the overwhelming relationship with music technology. But, as we all know, diving into the caves of our negative feelings is precisely where we do not want to go. Therefore, instead of looking for solutions from ourselves, sometimes it is easier to rely on the gadgets of the audio world that promise faster results and hit songs. But no tool, DAW (Digital Audio Workstation), or MIDI controller can protect us from the inner saboteur, which is always actively at work, ensuring that whatever we create, we will feel insecure in the process.

Yet, I feel relieved that I am not the only one feeling all this. As much as it might sometimes feel like we are alone with our feelings, this is a reminder that we all have inner saboteurs, and we all have the capability to tame them.

In this chapter, we will look deeper into creativity and confidence, and how they affect our practices as artists. We will start by looking into the big picture, which is demonstrated through a concept called *The Quartet of Creativity*. After this, we will discuss creativity on a deeper and more scientific level, to understand flow and inspiration. Last, there will be a conversation about creativity and what creative confidence means.

## The Quartet of Creativity

I am not a psychologist, sociologist, or a neuroscientist. I am an artist who fell in love with technology and its endless possibilities, and who feels passionate about figuring out what is going on in our mind while creating with these modern tools. On my journey into music production and making it my career, I have noticed the need in audio communities to openly discuss our insecurities about our creativity, work processes, workflows, tools we use, and our creative work's outcome and purpose.

This need to discuss artistic process in music production is currently very technical, only slightly touching on the reasons behind the issues leading us to have the insecurities around the art we make and how we make it. That is why, in this book, I am attempting to find out why we feel insecure, what are the most common excuses we give to ourselves to hide behind our anxiety and the ways we can overcome these issues.

In my research for these topics, I have concluded that to understand creative confidence and the insecurities within it fully, it is necessary to cover this topic through several different sciences. This book will bring together knowledge from people representing these scientific professions, personal experiences, and real-life stories from the people in the industry.

Creativity, Confidence, and Music Production   7

But as I am not a scientist specialising in any of these practices, I have developed a way to structure, analyse, and visualise creative confidence in music production through a concept I call *The Quartet of Creativity* (Figure 1.1). It is a simple way to discuss these topics and make them approachable to a regular musician and music producer. It divides our psychological, neurological, sociological, and physical functions into four categories: *Heart, Brain, Space*, and *Action*.

- The *Heart* (Psychology) is where you will find your self-consciousness, self-confidence, self-esteem, and self-efficacy, as well as identity and creativity.
- The *Brain* (Neuroscience) is the command centre of all your thoughts and actions. This is the science of your existence, the physical qualities, genes, and brain waves.
- *Space* (Sociology) is where all the previous characters exist. The world we are part of and which we experience with our understanding, and the society, culture, the music industry we try to navigate, as well as the social narratives of others.
- *Action* (physical attributes, haptics and interaction) is the deliberate effort you make to achieve goals and generate pleasure feedback, such as practicing, executing your plans, and moving your body to find inspiration or physically interacting with industry, communities, and physical instruments.

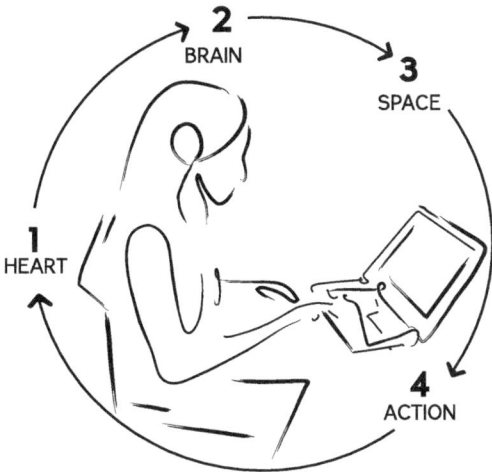

FIGURE 1.1  Illustrated by Emma Holdway

All of these four points are deeply connected and work together to make us breathe, function, develop, and create. The interaction and balance between these four points can therefore help us on our journey to identify the cause of the insecurities we have as musicians, artists and music producers.

> ### CASE STUDY 1
>
> To put *The Quartet of Creativity* into a practical context, let me introduce the first case study called Amanda. She is a musician who has been practising music production for around four years in total. Amanda dreams about a career as a professional music producer. Her biggest obstacle is to finish a whole song production by herself. And because she has never done this, she has constant doubts about her ever being good enough to do it professionally.
>
> So, let us break Amanda's story into the sections of *The Quartet of Creativity*. As mentioned above, all these four aspects of *The Quartet of Creativity* affect each other. Nevertheless, for clarity, we will now talk about them chronologically in a circular pattern, with the order: *Heart*, *Brain*, *Space*, *Action*, which all then feeds back to the *Heart* (as shown in Figure 1.1).
>
> Currently, she is creating a catalogue of tracks she can then use as an example to get her first paid job in the industry. The *Heart* is where everything starts. That is where her first inspiration, passion, and idea come for a song. The project's information is sent to the *Brain*, which then works as the manager of this whole process and where the action plan is created.
>
> Amanda's *Brain* gives the idea the go ahead. She sits by her favourite DAW to start working on the song. But, before the *Action*, she needs to conquer the most challenging phase of the project: The *Space*. Suddenly the idea that started from the *Heart* does not seem like such a good idea after all. Amanda looks at her DAW and remembers all the functions she still does not know, how her friends would disapprove of the genre she would like to make and how she is now wasting time on something that will never become a real, sustainable job.
>
> Nevertheless, she proceeds regardless and starts the *Action*. Two hours later, she is still trying to choose which synthesiser sound suits best for this genre and what drum sounds are correct to create the feeling she is looking for. Amanda feels tired and decides to continue the project the next day. The events of this whole two hours are now

fed back to the *Heart*, where the inspiration lies in the remnants of everything that the experience made her feel: success and failure.

However, the *Heart* does not only carry the emotional weight of this particular event. The *Heart* and the *Brain* include all the data of all experiences since Amanda's birth. The loop of *The Quartet of Creativity* has existed all her life, forming her confidence and identity. But like Amanda, we all have what it takes to look into what makes any of these four areas of *The Quartet of Creativity* block the flow to our best creative potential.

## Creativity and Music Production

When researching creativity, the concept can be very complicated and confusing. There are several significant scientific theories on why we are creative and its true definition. But at the same time, if we started a conversation with someone in a pub about creativity, all of us would most likely relate and understand it on some fundamental human level. Creativity is something we recognise and understand because we have all experienced it. But this changes when we try to define it, as it is a personal story about our childhood, how we grew to take chances and say what is on our mind.

When watching a TED talk about creativity by Sir Ken Robinson, it is hard not to relate to what he says. He talks about education and creativity in children:

> Kids will take a chance. If they don't know, they'll have a go. They're not frightened to be wrong. Now I don't say being wrong is the same thing as being creative. What we do know is, if you're not prepared to be wrong, you'll never come up with anything original. And by the time they get to be adults, most kids have lost that capacity. They have become frightened of being wrong. And we run our companies like this. We stigmatise mistakes. And we're now running national education systems where mistakes are the worst things you can make. And the result is that we are educating people out of their creative capacities.[1]

In this speech, Sir Ken Robinson catches the exact point of what inspired me to write this book in the first place. We all are unique, powerful, and complex as humans, but since we were children, we were made to fit in and care what others think of us. So when we enter schools, workplaces, and other social structures, we lose the confidence that our unique views and ideas are enough. That is why, even if we cannot all understand creativity scientifically, we can learn to believe that the way we understand and express it is what matters.

But when trying to define creativity, it can be useful to think that we can find creativity from the crossroads of our perspective and how we perceive the word around us, and our experiences and the knowledge we have learned[2] (refer to Figure 1.2). It means that all of our creativity is special and unique to us. This is why we also need to have more trust in our visions, as our past, our knowledge, and our view of the world, make them rare and extremely valuable. That is why we need the confidence to believe that our perspective matters and should be heard. But as the school system has taught us from a young age to question individuality and instead encourages us to fit in, we start to fear our voice and sensor what we say or create publicly.

As we take a deeper look at creativity's reasons and origins, we need to recognise the scale of its meaning. Some call it a phenomenon, a skill, a tendency, or "out of the box" thinking. Others believe it is connected to intelligence and can be measured, while others see it as an intangible part of our human capacity. How you study creativity and how you experience it depends on the filter of your looking glass.

It is easy to think that there are rules to creativity, especially in music production. Most of the insecurities in this book are somehow connected to the idea that there is a right or wrong way to make music. I always like telling my students, "Only science is right, and the rest is just opinions". It means that how LFO (Low Frequency Oscillator) makes a sound wobble

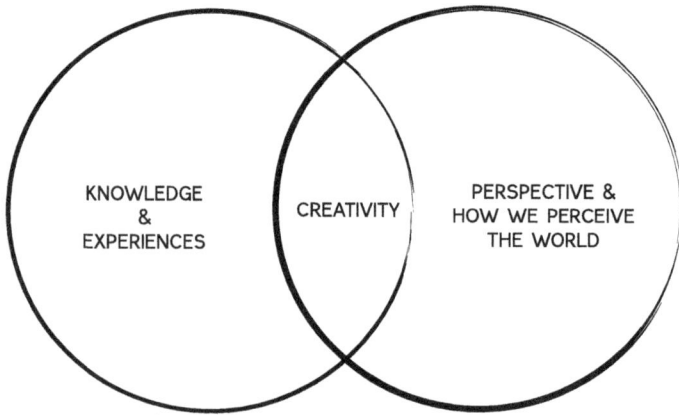

FIGURE 1.2 Illustrated by Emma Holdway

is to do with its acoustics, the study of sound, but how you apply LFO to the signal is your creativity in action. But as the music and audio industries evaluate art by genres, rules and guidelines, how can you know how to focus on your original ideas? Therefore, when it comes to music production, we should separate our technical knowledge and personal creativity into two separate sections, giving them equal attention.

What can cause conflict with separating knowledge and creativity is the thought that creativity and intelligence are connected. Psychologist J. P. Guilford introduced the idea that creativity can be measured via dedicated tests. According to him, the more creative potential and performance one has, the more intelligence one possesses.[3] On the other hand, studies have shown that creativity is a quality that we all have from birth. For example, Robert Weisberg conducted years of research defining creativity as problem-solving.[4] The idea being that each project or strand of inspiration is a puzzle to be solved, meaning that no matter how big or small the task is, it can be valued as creativity. This eliminates the notion that intelligence plays a part in creative practice. Anyone can be capable of solving a problem that has been presented externally or by your own imagination, and for this you need creativity.

In a way, both of these perspectives can have value depending on the goal, experience, and perspective an individual has on their personal creativity and creative identity. We all know people who have traumas from school art classes, where drawings were compared and measured by our technicality, leaving us to believe either that we are not creative individuals or that we are creative geniuses. The values we grow up with and the feedback from experiences can make us believe that creativity is intellectually connected or that it is more of a *talent* that only some people possess. Alternatively, creativity has become increasingly valued in the business world, manifesting in staff training on brainstorming, design, and innovative thinking, leaving behind the times where creativity was only for artists and people with specialised job titles.

Similarly, some consider that intelligence is a big part of learning STEM topics: science, technology, engineering, and mathematics. So, we live in a world that can make us believe that only certain people can be creative, that creativity can be measured and that you need an extraordinary level of intelligence to learn technical skills. It explains why so many of us perceive right and wrong in the music production, audio techniques, and outcomes of the track.

As a music producer, I feel overwhelmed by the science of creativity. Still, I find it highly beneficial to know what contributes to the aspects of my mind that can help me achieve my artistic, life, or career goals. Our creativity is a powerful tool that should not be taken for granted. But instead, we need to focus on believing in it, nourishing it and taking care of it.

## Confidence

Confidence is a popular word that is widely used to describe a person's belief in themselves. It is a broad and confusing concept to understand, as the term can commonly mean almost anything depending on the person's perspective. To better understand what confidence is, it can help to understand the differences between *Self-Efficacy*, *Self-Confidence*, and *Self-Esteem*.[5]

- *Self-Efficacy* is believing in yourself and your ability to do well and achieve your goals. It is about having a positive outlook on your future and feeling confident you can handle whatever comes your way.
- *Self-Esteem* is all about how you see yourself right now. When you have a sense of self-worth and feel good about yourself, this includes how you think about yourself and how others see you.
- *Self-Confidence* is about trusting in your abilities and skills. It is about firmly believing that you can do what you set out to do. Self-confidence can help you feel in control of your life, set realistic goals, and understand your strengths and areas for improvement.

When we experience low self-esteem, we tend to close up and focus on the negative feelings that the situation brings up. Maybe you feel like you are not good enough or that you are letting someone down. Due to low self-confidence, experiencing these feelings can prevent you from achieving things you would like to do and slowly limits the scale of your dreams, and how you explore and develop your skills. All these emotions are common, and we all feel them frequently. And for some, it can be helpful to seek professional help to navigate our complex brains and find balance.

Fear is additionally one of the most significant contributors to our insecurities. Most of the time, we might be unaware of our fear and how much it influences our actions and emotions. We can feel it in our everyday lives, especially when we break our routines and step outside of our comfort zone. We can experience triggers from our past experiences that make us feel negative feelings, such as shame and hurt.[6] Fear and fear of failure are massively connected to many topics discussed in this book. You can read more in-depth conversations about how it affects our creative confidence from Chapter 3.12.

Fear is also a factor to the stress we might experience around music-making, publishing, and generally being part of the music or audio industries. How we feel in ourselves directs the ways we wish to put our visions into action and in what way we will feel the most comfortable to practise our creativity. This is why it is essential to mention substance abuse as an important part of the discussion on creative confidence. Music therapist Rachel Jepson mentions in an interview how drugs and alcohol is often a topic within the music community:

> You can't talk about addiction without talking about stress, the music industry and confidence. They are all completely tied in. Some people feel they need to be drunk or take some substance to be confident and create. Many well-known artists and bands have been drunk or wasted on drugs when writing great music. And then there's a myth that you have to be under the influence of alcohol or drugs if you will write music and do well.[7]

Another part of this conversation is mental health and how navigating balance between the music industry, artistry, and our own well-being can be difficult. In our cultures, we value authenticity in art. Still, sometimes it might come at the expense of an artist, who might put themselves in a vulnerable and exposed position by publicly opening up about a painful past or mental health. In one way, it can be inspiring and helpful for others and a great way to express feelings, but at the same time, it can also make it difficult for the artist to find balance in their life. Jepson continues:

> Creative people often struggle with their mental health, and they are predisposed to struggling with depression and anxiety. Often that's worked into their music. Although it is brilliant to have that outlet, it can also be challenging to navigate. If you're writing something and you might feel depressed, it might be cathartic to write down what you experience. But is you or your label making you do this? There is so much within that, which can be complicated.[8]

Even though mental health and substance abuse is an important conversation to have related to creative confidence, these are very big topics to cover and are too wide to discuss in the detail that they deserve, in this book. However, at the end of this chapter you can find additional resources if you wish to look into these topics in more depth.

> ## PSYCHOLOGIST COMMENT
>
> An extract from an interview with Tricia Greenwood.[9]
>
> > Many of the beliefs that cause us to feel inadequate or inferior stem from negative patterns of thinking. Some of the most

common corrosive tendencies are comparing ourselves to others, having an excessive need for approval from others, and judging our worth through our achievements.

Our *inner critic* often focuses on the parts we do not like about ourselves, and we assume that others judge us as severely as we judge ourselves. The *inner critic* is often harsh and unforgiving and drowns out any positive and affirming thoughts. Choosing to focus more on our qualities and strengths helps to foster a more balanced and healthy self-image. Challenging long-held negative beliefs about ourselves can help to redress the thinking that causes us to feel bad about ourselves.

When our self-confidence and esteem are under attack from our thoughts – our evidence for reaching skewed conclusions often emanates from our perception of how others view us. When someone blanks us as we wave at them, we might assume that they are deliberately ignoring us when in fact, they genuinely did not see us. By checking out these sorts of inaccurate assumptions, we can start to counter the negative thoughts we may have about ourselves with more balanced and realistic ones.

When our self-efficacy – our belief in our ability to achieve – is low, we lose the confidence and the motivation to try or persevere, which in turn becomes a significant obstacle to succeeding. Our self-efficacy is influenced by the things we have been able to achieve in the past. If we experience repeated failure, it can diminish our efficacy and discourage us from trying again. Not achieving what we want may well be associated with low self-efficacy rather than a lack of talent.

## Insecurities

Our creative confidence is fragile. It can be affected by anything, especially when we feel failure. Our perfectionism, jealousy, envy, and the feeling that we will never be good enough are based on fear of failure. But by redefining this word and learning to believe that there is no failure in creativity, we can start to use our insecurities as a map to change what is hurtful to us. By understanding where our insecurities come from, we can also start to understand other people better, compare each other less and be kinder to ourselves and others.

All of our insecurities come from somewhere, we were not born doubting ourselves, but a combination of factors from your childhood and life have contributed to the self-hood you have at this current time. Who we are born as and who we will become are an eternal cycle of influence from others and the world. As Svenja Weber and Gianpiero Petriglieri say about insecurity:

> Insecurity is a *social* issue with psychological consequences, not a *psychological* issue with social consequences.[10]

It can be challenging to differentiate which aspects of our capabilities and thought processes are affected by society and environment, in contrast to what we were physically born with. For some, it might be easier to believe that we are wired a certain way at birth, as it can bring comfort in distress, particularly when things are not going as hoped. In these cases it is easier to find explanations from things we believe we have no control over, such as "I am not capable to do this as I am just not built this way". But by finding awareness in what social aspect has affected our psychological qualities, can reduce self-blame leaving us space to re-evaluate and grow in the direction we desire.

We all might have different beliefs about the nurture vs nature argument. Some might have religious and faith-related beliefs about what controls our actions and feelings. Some turn to evolutionary science and form their opinions on how we act based on that. Then there are all the different sciences, such as neuroscience, psychology, and sociology, that might have different takes on this conversation. However you believe you were made or what controls you, the key question is whether you think you or another human have the ability to change.

### Author's Experience

When I think about what aspects of my life have caused me the most insecurities, I think of three significant points. All these events in my life were moments where I needed to challenge my beliefs about whether I could change or not. Being diagnosed as neurodivergent from a young age, being bullied throughout my school years and later having my first negative music industry experiences before knowing my strengths. There are plenty of other influencing factors, but even recognising these significant moments in my life can help me break down how they affect my current state of creative confidence.

For example, being diagnosed with dyslexia and dyscalculia as a child gave me the message that neither academic or traditional success will be expected from me. Therefore, low expectations led me to seek validation through activities other than studying, as I did not see the

> point in getting good grades in school. If the world is not built for me, why would I need to try to fit in it.
> 
> This led to decades of insecurities, believing there was something wrong with my brain and that I was more stupid than others. This was until I independently learned more about dyslexia and dyscalculia and understood that I had only been taught the negative meanings of these labels. The system had failed to inform me about the positives, which made dyslexia almost a superpower. For example, I can be extremely good at holistic thinking, sound recognition, problem-solving, spatial awareness, and even geometry.[11]
> 
> This realisation enabled me to change my self-identity and how I saw myself as a learner. I forgave myself for ever doubting my abilities and understood it was not my fault that I had felt this rejection for so long and that there was nothing wrong with my brain. Still, there are plenty of dangerous thought models within our society that had made me think that the fault was mine. I studied two degrees, using a foreign language, graduated at the top of my class, with a master's in music production. I had realised that it was up to me to take responsibility for my success.
> 
> Similarly, being bullied made me care what others think and taught me to seek validation. A bad experience in my first music industry job made me question my capabilities as a songwriter and a singer. Now, in my thirties, I am trying my hardest to be kind to myself, to understand where these negative thoughts come from and constantly remind myself that I am in charge of my identity and narrative. It is up to me to leave the old harmful thoughts behind and recognise my capability to find my confidence again.

Whatever you might have experienced in your life and however the world around you is currently making you feel, the narrative about yourself and who you wish to be is only up to you. It is about you believing that your ideas and views are valid and unique. You are entitled to keep your creativity private or share it with the world, but the value of the creative act, process or outcome, is defined by you.

## Searching for Flow

Creativity and the concept of *flow* are relatively new ideas in our history. But like creativity, we all know what the word means and what it possibly feels like. Consequently, studies on flow often fail to provide conclusive answers and instead prompt us to ask more questions about their possible causes and how we can control our own creativity.[12] Flow is a highly personal experience influenced by numerous factors, making it impossible to provide definitive answers on when and how we achieve creative flow. But on a personal level, we can learn to understand how and where our creativity happens, and how the flow-state gets the best chance to be activated.

I have figured out that structure, time schedules, and limitations help me immensely in finding inspiration and flow. My head often feels so busy with ideas and thoughts about what I could create, which, together with several instrument and plugin options, can make me feel overstimulated and overwhelmed. This often ends with me getting exhausted and wanting to walk away from my creativity. For others, structure can create pressure to achieve, or they want lots of possible tools to allow flexibility in their creative approach. I have only experienced creativity and the feeling of flow through my own perspective; therefore, this book and the techniques in it come from what I feel has worked for me and my students. But I hope these thoughts and perspectives allow you to reflect and find whatever works for you.

In a discussion, the professor (PhD in Behavioural Neuroscience), sound engineer, and record producer Susan Rogers talks about how often creativity can be found by turning off the logic in our heads and finding a way to activate the more intuitive side of our mind:

> What we have to do is activate regions of the brain that are involved with feeling, sensing, postponing and deciding. That feels like going into our own heads and shutting out the outside world, but that's really kind of what happens: we do go into our own heads. And when we activate the part of the brain that is "in our own heads," we're activating circuits that also get busy when we're creative. So, in a sense, we go back to being almost like a little kid where we don't have rational thoughts so much, where we don't have tasks and obligations. We're not looking forward into the future, we're not looking backward into the past. We're in the moment.
>
> And that's a little bit, for some people, scary because you have to shut off what's coming to you. You go to a very private place. Very creative people can do that very quickly, very easily. Very easily. It's like a muscle, and the more you exercise it, the better you get.[13]

The challenge can be firstly to recognise whether our process is led by intuitive or logical thoughts, but also how we can approach the more emotional side of us without fears. One theory that was a game-changer for me in recognising how my own creative process works was becoming familiar with the critical modes of thinking originally introduced by J. P. Guilford: *divergent* and *convergent thinking*.[14] Divergent thinking is where we reach for new ideas and let our imagination fly without judgement or systematic reasoning. With convergent thinking, we analyse and criticise the ideas we have discovered. This is the stage at which we use the material created in divergent thinking to form an answer to our creative puzzle.

In music production, our workflow often includes several different practices at the same time: brainstorming, learning, technological knowledge, musical artistry, critique, and constant comparison to existing musical work. Therefore, we often try to use divergent and convergent thinking and do learning simultaneously, which can lead to confusion and trigger insecurity. After pulling out new ideas, allowing ourselves to judge them straight away creates space for our insecurities to overpower our creative confidence. But doing all these processes separately can give us surprisingly positive results in our creative outcomes.

Yet, it seems impossible in music-making to separate our workflows into creative thinking, puzzle solving, and learning, as most of us are hoping for the deep concentration where all of these three happen together in complete harmony. In arts, we often refer to this state as *flow*. It describes a moment of total focus, in which you feel nothing but inspiration to continue creating with constant pleasure feedback and a non-judgemental mentality. As musicians, we desire to have a flow state every time we sit by our instrument or tool. Still, often the moment gets cut short by one of the many contributing factors: distraction, insecurity, confidence, fatigue, time, or feeling overwhelmed.

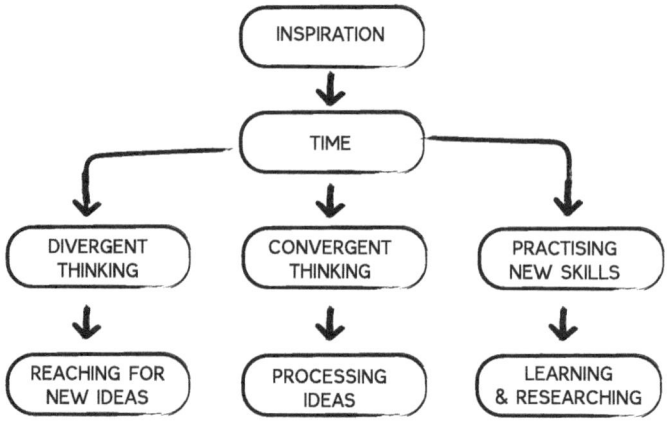

FIGURE 1.3 Illustrated by Emma Holdway

Especially for music producers, these distractions significantly contribute to losing the flow of inspiration. Interruptions can arise in the form of technical issues, equipment limitations, or gaps in knowledge, ultimately impacting our creative confidence. When the sound is not coming through, it can make us feel inadequate, and strong emotions like shame or disappointment can shift our focus away from creative inspiration, thereby disrupting our flow. Later in this book, we will further discuss how these exact reasons can interfere with our flow and how our confidence impacts our creative workflow. As mentioned, these topics are vast and still not well understood. Therefore, this book explores them primarily through my own experiences and observations in the music industry.

When starting to write this book, I was determined to find answers from science that would help us find a flow-state, but after writing this chapter many times, I was faced with the reality that you cannot force it to happen. There are many theories and ideas on how we possibly could reach it by practice, willpower, or other techniques, such as meditation. For example, the Harvard Health article talks about meditation in a way that sounds very similar to a flow-state:

> There is more than one way to practice mindfulness, but the goal of any mindfulness technique is to achieve a state of alert, focused relaxation by deliberately paying attention to thoughts and sensations without judgment. This allows the mind to refocus on the present moment.[15]

This can be a very beneficial approach if you are a person that struggles to take a moment for yourself in the middle of a busy life and you need headspace to give you a moment with your creativity. But in the end, the benefits of meditation in the search for a flow-state depend so much on the person and how someone is using it to approach their creativity. Therefore, just as there is no right or wrong way to be creative, there is no perfect way to find our flow either. We need to feel comfortable with the idea that flow is impossible to harness into a formula that we can use every time we wish to experience it.

In an interview, Susan Rogers continues about the impossibility of forcing flow in our creative practices but emphasises how consistent practice and play, even on the days when you might not be feeling it, increases the chances to find inspiration and a flow-state:

> We shouldn't think of creativity as something that we either have or lack. We have all got it to some degree. I love how my friend, the

sculptor Tim Bruckner, says, "Craft is what sustains you when art fails you, which it will 90% of the time." Art is original thought, art is inspiration – that's your creative self. For the greatest artists, it's just 90% of the time not there.

So what these artists do, whether it's Ernest Hemingway going to his desk every day whether he has an idea or not, or the scientist going into the lab and looking at data, or the musician going into the studio and just getting the drum groove going (Prince did that lots of times), you're just practising your craft. As you practise your craft – not always, not even Prince, and he was hyper-creative – occasionally everything will just kind of line up, and creativity will flow in that moment, and you'll get an idea.

You have to be comfortable with the notion that you can't turn this on like going to a faucet and saying, "Okay, come now, I'm ready." It will not behave like that. By "it," I mean your brain. So what you should always be doing is practising your craft. The other thing you want to do is get out there in the world and observe. Have fun and look at other people's creativity. And quite often, that is going to inspire your own creativity.[16]

The idea that we cannot force flow or inspiration can feel rather frustrating. But in the end, I do think we can invent custom rules, techniques, and

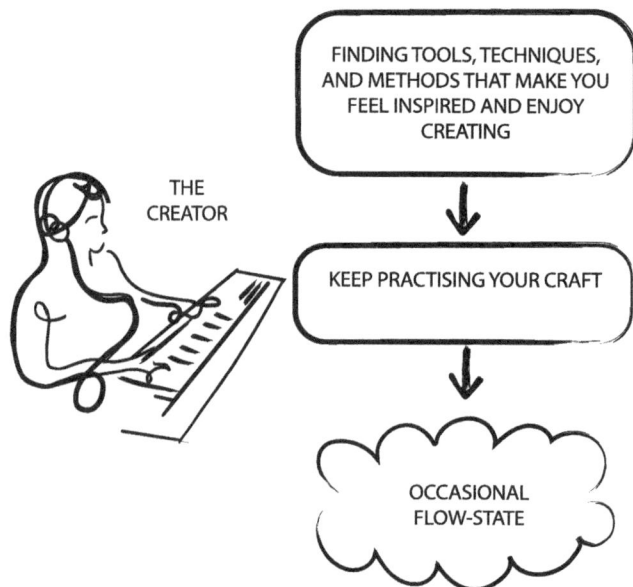

FIGURE 1.4 Illustrated by Emma Holdway

guidelines that help us to be more consistent in our creative practices and help us manage the obstacles that would otherwise stop us from focusing on our craft. And I think that is one big reason I wanted to write this book in the first place, as I can see this in myself and others around us. We feel pressure to feel inspired and creative. But before we can do that, we struggle even to start or to be fully present with our craft. Everything you will read in this book and all the practical tips are focused on helping you find what works for you. After something feels like it is making you enjoy your creative moments, you keep on repeating it, expanding, and exploring. And I am sure the surprise visits from your flow state will become more frequent.

In Chapter 4, there will be further discussion and practical tips to put these ideas into practice. We will explore how to understand, practice, and implement different thinking methods in our everyday creative practices. They will help you to create both life habits and workflows that will allow you to focus on what you want to achieve and prevent your insecurities from taking over.

In the next chapter (Chapter 2), we will discuss confidence and creativity concerning technology. We will further examine what aspects of the music and audio industries, from the interface design to music spaces, affect our insecurities and creative potential.

## Notes

1 Sir Ken Robinson. 2006. "Do schools kill creativity?" YouTube, uploaded by TED, 7 July 2009, https://youtu.be/iG9CE55wbtY.
2 Developing the Cambridge learner attributes. n.d. "Chapter 4: Innovation and creativity". *Cambridge International*, 22.
3 J. P. Guilford. (1 October) 1966. "Measurement and creativity". *Theory Into Practice* 5(4): 185–189.
4 Robert W. Weisberg. 1993. *Creativity: Beyond the Myth of Genius*. WH Freeman.
5 Courtney E. Ackerman, MA. 2021. "What is self-confidence? + 9 ways to increase it". Accessed 15 January 2024. https://positivepsychology.com/what-is-self-confidence/.
6 Courtney E. Ackerman, MA. 2021. "What is self-confidence? + 9 ways to increase it". Accessed 15 January 2024. https://positivepsychology.com/what-is-self-confidence/.
7 Rachel Jepson. (March) 2022. Interviewed by Liina Turtonen.
8 Rachel Jepson. (March) 2022. Interviewed by Liina Turtonen.
9 Tricia Greenwood. (September) 2021. Interviewed by Liina Turtonen.
10 Svenja Weber and Gianpiero Petriglieri. (27 June) 2018. "To overcome your insecurity, recognize where it really comes from". *Harvard Business Review*.

11  Matthew H. Schneps. 2021. "The advantages of dyslexia". *Scientific American.* Accessed 14 October 2021. www.scientificamerican.com/article/dyslexia-can-deliver-benefits/.
12  L. Harmat, Anderson F. Ørsted, F. Ullén, J. Wright, and G. Sadlo, eds. 2016. *Flow Experience: Empirical Research and Applications.* Springer.
13  Susan Rogers. (July) 2024. Interviewed by Liina Turtonen.
14  Harvard Professional Development. 2016. *Convergent vs. Divergent Thinking.*
15  Harvard Health. n.d. "Benefits of mindfulness". HelpGuide.Org.
16  Susan Rogers. (July) 2024. Interviewed by Liina Turtonen.

## Additional Resources

- Csikszentmihalyi, M. 1996. *Flow: The Psychology of Optimal Experience.* Harper & Row.
- Eagleman, D. 2017. *The Brain: The History of You.* Vintage.
- Eagleman, D. and A. Brandt. 2017. *The Runaway Species: How Human Creativity Remakes the World.* Catapult.
- Gilbert, E. 2015. *Big Magic: How to Live a Creative Life, and Let Go of Your Fear.* Bloomsbury Publishing.
- Green, B. and W. T. Gallwey. 1986. *The Inner Game of Music.* Doubleday Books.
- Kahoud, D. and D. Knafo. 2018. *Sex, Drugs and Creativity: Searching for Magic in a Disenchanted World.* Routledge.
- Robinson, K. 2009. "Do schools kill creativity". YouTube, uploaded by TED, 7 July 2009, https://youtu.be/iG9CE55wbtY.
- Shetty, J. 2020. *Think Like a Monk: The Secret of How to Harness the Power of Positivity and Be Happy Now.* Thorsons.
- Smith, J. 2022. *Why Has Nobody Told Me This Before?* Michael Joseph.

# Creativity, Confidence, and Technology

**2**

## Keywords

- STEM
- STEM identity
- interface
- plugin
- haptic interaction
- feedback loop
- imposter syndrome
- modular synthesiser
- MIDI
- MIDI controller

## Our Relationship with Technology

What is your relationship with technology and how does being around it make you feel?

The answer to these questions might look very different depending on what generation you were born in. Whether you used the internet for the first time as an adult, and remember the sound of a phone modem or had your first smartphone at the age of 5, your relationship with technology is unique in that it has the capacity to intimidate you, challenge you or empower you.

### Author's Experience

I am a millennial who grew up during the years of the most significant shift in technology. In Finland during the 1990s, where Nokia was ahead of the curve, and graphical internet browsers were invented.

DOI: 10.4324/9781003194484-3

As much as the developing technology was around me both at home and school, I was still surrounded by a society that made a career in technology seem very exclusive – a possibility only for those succeeding in school, especially in mathematics and sciences (STEM subjects).

Growing up, I was a social and outgoing girl with severe dyslexia. I was never expected to do well in STEM subjects, and a future in a technological industry was not a possibility. Instead, I was directed towards creative subjects, as those were the easiest for me. After a while, this became part of my identity. I was proudly telling jokes about failing all mathematic courses at school and as we know, joking about insecurities is often a defence mechanism. I took the role of someone with a learning disability and prepared surviving in the world without fitting in.

When I was around 12 years old, I remember finding my dad's old computer in the back of a closet. I took it into my room and pieced it together, cable by cable. When it worked, and I managed to get my favourite game up and running, I felt this incredible empowerment. I was as knowledgeable and nerdy as my big brothers were, and this made me feel amazing. Looking back, it was my way of finding a connection and a personal relationship with technology. I saw it as a tool for power, knowledge, and validation, something I wanted to be part of but did not know how to be.

My experiences have made me question everything we are taught about technological talent, intelligence, and the role of creativity in it. Finding music production gave me a STEM identity. Without it, I would have continued as an acoustic singer-songwriter for the rest of my life, relying on others to do the technical work, not knowing the millions of other ways I could express my musical creativity as an artist and an engineer.

Our relationship with technology and how we identify with it have a massive impact on our creative work. If technology is so available to us, why do more people not use it as a tool? Why are we scared of it? Why do we keep technological audio knowledge at such high value, making us feel insecure about our artistry?

This chapter will look into these questions further and discuss how technology affects our creativity. Especially as music producers, we can find ourselves at unusual crossroads of several different industries, such as art, music, audio, and technology. These topics can seem very detached from each other, but in the end, the conversation always comes back to our confidence and insecurities. When a computer is your primary tool to make your vision come to life, how does it affect our creative workflows, and what part does confidence play in all of this?

## Technology and Music Production

The definition of a music producer has changed drastically through recent years. For most of my life, I never really gave any thought to the people in the studio. The engineers and producers were these distant characters that lurked in the shadows, and personally, there was nothing about these people I could relate to. Not until I saw the music video to the song "Uptown Funk" by Mark Ronson, featuring Bruno Mars.[1] This song brought a producer out from the shadow to my screen with flashy outfits and star quality. For the first time, I understood that producers could be artists with their own music and unique style. In the past decades, Mark Ronson is just one of the many producers making their brand as an artist in the music industry, inspiring many aspiring producers along the way and sculpting the way we understand this job title in the 21st century.

The evolving description of music producers and the constantly developing technology has also changed the *where* and *how* we create music. You do not need to go to big studios with expensive equipment to create hit songs anymore; everything you need for making a professional-quality recording is now available for anyone with a laptop or even a phone. The industry has gotten used to the concept of a *bedroom producer*, which used to have a slight undermining connotation, but now is a regular job title. The new generations of pop stars, DJs, and artists can be developed in the privacy of their homes, proving that anyone can be the latest big name.

And, it is not just the music industry that is benefitting from these advances. It is also having an effect on the way that audio hobbyists, music therapists, and sound designers for film, games, art, dance, and theatre are working with audio equipment. The constant development is making it more accessible to people without degrees in audio who may ordinarily have to rely on external engineers. These tools are for regular people to use, and this is changing how we interact with technology, how we create with it, and how it will develop in the future.

We are used to having our phones, laptops, and other daily electronic equipment as part of our identities, routines, and social lives. They are accessible, easy to use, and allow us to learn new skills such as photography, video editing, drawing, and music-making with just a few clicks.

In part, technology has found a way to coexist in our lives in harmony, but elements of this relationship still make us feel confused and insecure, particularly when it comes to our creativity. It is as if our societal expectations and the development of technology have not grown at the same pace, and this is the reason why STEM subjects are still out of reach to big marginalised groups. Using these modern electronic instruments can make us prioritise knowledge and techniques over artistry, and this is why we need so much confidence to learn technology.

As a result of the way the music industry used to be, we still think there is more value in a song produced in an expensive music studio than one made with a free app on a laptop. Furthermore, as artists, we often feel like we cannot call ourselves professional producers if we do not know the ins and outs of all our equipment or we do not have a formal education in the subject. Therefore, why do we judge the art we make: by the technology we use or the value that the industry puts on specific knowledge? In the end, if the song makes us feel something, should that not be the only thing that matters? Defining art by an experience, not by the price of a microphone?

## Approachability and Accessibility

Approachability and accessibility are some of the key concepts in modern audio design. They affect our relationship with electronic music instruments from how we first get introduced to them, how we feel using them, how we create with them for the first time, and how we continue developing our skills with them.

I grew up in the classical music world, playing the violin in a conservatoire from the age of five, spending evenings in choirs or orchestras. But I never felt like I fitted into this environment, with its rules and grading procedures, and a sense of elitism in it. Finding music production and electronic music gave me new freedom and a confidence in music which I had never felt before. Suddenly the only rules were determined by physics and how audio works, not by humans. I could take my violin, loop it, distort it, explore sounds however I wished to and do it on my terms, without concentrating on finger techniques.

But saying this, I recognise that there is beauty in mastering an acoustic instrument, such as the violin, using rules as guidelines for finding the individual artistic expression from the violin's tonal capacity. That is why I do not think we should evaluate and contrast learning processes and musical styles by putting them in hierarchies or defining their value. Instead, we can learn to value all musical instruments equally and give them a chance to cooperate in harmony. If anything, removing these borders or making them more flexible can only enhance our creativity and encourage new musical directions.

Alternatively, it can be argued that the vast amount of electronic instruments, their value, and technological challenges do not encourage the user to master musicianship of one particular device, therefore seeing them as somewhat disposable. Musicians might buy a piece of equipment, feel inspired by it, and when another shiny new tool comes to the market, the old one is left in the cupboard to collect dust. In an interview Marie Tricaud, an interaction designer, commented on the comparison of acoustic and electronic instruments:

> For centuries before electronic music, something like the violin has been refined for hundreds of years, perfecting this one instrument. Acoustic instruments are shaped after the sound that they produce physically. Whereas electronic music is made with instruments that synthesise sound electrically, electronically or digitally, and so suddenly the design possibilities of their shapes and/or interfaces are endless. This example of contrasting violin and electronic instruments explains the number of electronic instrument interfaces that exist, as well as why it might be interesting to switch between these tools because each one gives you a different way of thinking and exploring creativity.[2]

Tricaud also pointed out how a new student approaching the violin can have a completely different experience from approaching a modular synthesiser. Most people know what sounds to expect from a violin, understanding that it takes practice to achieve specific sounds. Whereas with a modular synthesiser, it is almost impossible for a beginner to understand and recognise the sounds that come from it. To learn it, you would need to know the technicalities of the components, how to connect them, and a basic understanding of audio physics. Also, the approach to modular synthesis would depend on how the tool looks physically so that we have the confidence to start figuring out all of its secrets.

For example, the Vochlea Dubler Studio kit is a device that converts voice to MIDI. The technology of this tool is complicated, allowing this MIDI controller to receive an audio signal and feed it as data to a DAW. However, from the outside, it looks like any other dynamic microphone. The familiar-looking interface takes away the user's initial feeling of intimidation, which advanced technology could alternatively create. From a psychological point of view, this can make the user feel calm, as they know nothing extra is expected from them, no audio physics knowledge or degree in engineering.

In contrast, developers can create tools that purposely do not hide their technological components. By designing an interface that looks more complex, a developer can reach a specific customer base and potentially sell the instruments at a higher price. In some cases, the approachability and accessibility of an instrument will be overlooked in favour of a design that is more in keeping with a brand image. For example, a complex vintage look on a synthesiser can give the perception that an instrument is "more advanced" or of higher quality.

Conversely, a simple-looking and toy-like cover for a piece of powerful equipment can have the opposite effect. Is the tool worth the money if it looks uncomplicated? This question was asked when Teenage Engineering released their TX-6 stereo mixer to the market. How can something so small and innocent cost $1200? This is an excellent example of how we can have unconscious biases towards what we consider advanced and better quality. A complicated-looking interface might not be welcoming to people unfamiliar with its functions or purpose. They might not identify themselves as someone with the ability to use technology that looks as complicated as a modular synthesiser. But if the tool looks like the TX-6, anyone, regardless of the background, might feel comfortable approaching and trying it. But often, the target audience of these tools is not the regular consumer, but rather the people that the developers themselves can identify with. This can make it challenging to reduce the gap between people who do not recognise their STEM identity when using these tools and those who are comfortable in their relationships with technology.

Fast technological change has enabled music to be available to everyone. Audio equipment developers are constantly reaching for new markets, so that studios and individuals within the music industry are not the primary focus anymore. In the last decade, the need for interactive, affordable, and approachable audio equipment has become increasingly popular in gaming, education, music, and art therapy, and other art sectors. Improved accessibility tools have been added to DAWs such as Ableton Live 12 which works with screen readers and Braille displays. A bigger consumer market means more viable business revenues for the new instrument, tool, and interface developers. And within more sales, market research, and competition, we find ourselves living in exciting times where the audio industry, and how we create, can change faster than ever before.

It is thanks to the accessibility and approachability of modern technology that I can call music production my career. When I tried recording audio for the first time, I had tools in hand that were budget, fun to use, and non-intimidating. I had a borrowed analogue mixer, GarageBand software that came free with my Mac laptop, and a one-pound karaoke microphone from a pound shop. Creating with these tools in a bedroom made me feel safe exploring, making beats, and recording bad quality vocals. Through this experience, I had the confidence to step into a music studio in the first year of my bachelor's degree. And now as a professional, I can help others to build their STEM identities by showing that technology is not as difficult as it might seem. The tools are there to assist, challenge, inspire, and entertain us, regardless of our background.

This conversation of approachability and accessibility in interface design will be discussed more later in this chapter. It is also the core topic of the first insecurity discussed in the next chapter: "I just don't have a techy brain", which dives deeper into STEM identity and explains why some technical tools are a cause of anxiety and impossible to approach confidently.

## Creativity and Confidence: Interface Design and Interaction

From the influence of more users and technological development, there has been a considerable increase in a more user-focused design approach. Devices are no longer just external tools that serve a purpose recording audio or making it sound good, they have become an extension of our creativity. Musicians have started to look for the same expression, feedback, flow, haptic interaction and physical attributes in electronic instruments as they do in acoustic. Similarly, with a broader consumer market, approachable interface design has become an important focus, increasing users' confidence and limiting their self-consciousness when approaching new devices.

Within the technological development we have needed to start to reevaluate how we discuss and approach evaluating musical expression between acoustic and electronic instruments. For example, if we can hear a musician's expression behind a controller the way we recognise a pianist behind their instrument, does this change how we should look at music education? For some, comparing a guitar and a computer might make you uneasy and feel like we are losing skills to learn tools in the same way that we used to. But with the development of all instruments, we are not losing anything, we are only gaining. There might be more options for musicians to choose from and due to this, we seem currently to be in an adjustment period for the educational systems.

Now it is common for higher educational institutions offer music production and audio technology courses and modules for all music students. But does this mean music education is also changing for children? In the recent years schools have started to take big steps towards teaching MIDI controllers and music production tools alongside the traditional instruments such as guitar or piano. James Tuck, Ableton Certified trainer and an educator based in England, has been trying get recognition for pad-based controllers such as Push for live performances for GCSE. In an interview with him, he mentions how even though some schools already offer music production education there are still obstacles to get it standardised in the system. According to Tuck, some of these obstacles are reaching suitable exam board requirements, training the PGCE/trainee music teachers, cost of the equipment, support from IT department in school and keeping up with the changes and development this sector requires. Although it is clear there is still lots to do before music production could feature in qualifications such as GCSE Music through more relevant applications of music production compositional techniques, Tuck is positive that the conversation has started to make this happen and we might see some big changes in the future years.[3]

As we can see, the field of music education is constantly evolving and this development has brought about exciting changes in the music industry, especially in the area of interface design and modern music technology. The designers of the 21st century consider designing as a way of thinking, instead of concentrating entirely on a great product and the experience.[4]

Several different practices are applied in the invention process of designing one piece of audio or music-making equipment:

- *User experience* (UX) focuses on *why*, *how*, and *what* the product is used for. The aim is to cover all aspects, from the users' enjoyment to how they will troubleshoot the device.[5]
- *Design thinking* is a technique to find ideas, challenge them, and identify ways to test the products most appropriately.[6]
- *Human-Computer Interaction* (HCI) is the study of the relationship between users and computers.[7] Audio HCI practices have increased in recent years, focusing on the critical research areas of artistry, workflows, haptic interaction, and creativity between musicians and electronic instruments.

HCI, UX, and design thinking make it possible for us to explore new dimensions of creativity with tools that allow entirely new ways of interacting with sound, challenging our artistry and musicianship.

As a music maker and producer, it is exciting to see new ways to interact with our electronic instruments each year. In 2023, Ableton Live introduced fully MPE compatible pads on Ableton Push 3. As soon as it became

Creativity, Confidence, and Technology    31

available, I tried it out. It felt natural to add vibrato, slide between notes, and bend the pitch like I would do with the strings of a violin. This experience felt organic and familiar, yet unique due to the pad layout and how each synth sound interacted with the movement. Playing with colourful pads that feel good and react to my touch on multiple levels is highly satisfying and almost addictive.

This type of interaction is an essential part of HCI and User Experience (UX) development. When the controller is designed to provide a positive sensory feeling, it can influence the creative choices of the user and create a *haptic feedback loop* (see Figure 2.1). This interaction can then contribute to the flow-state we experience in our creative workflows. For example, when we touch the soft pads of a Push, we get a positive sensory feeling, and the lights, colours, and sound response that follow make us feel satisfied and excited. This interaction is similar to the basic principles that children feel when playing with toys. To make this exchange successful and enhance our creative workflow, it should challenge us just the right amount and bring enjoyment without making us think it is too challenging or overwhelming.

Engaging in a feedback loop repeatedly, without any interruptions, can boost our confidence in working with the tool and encourage us to

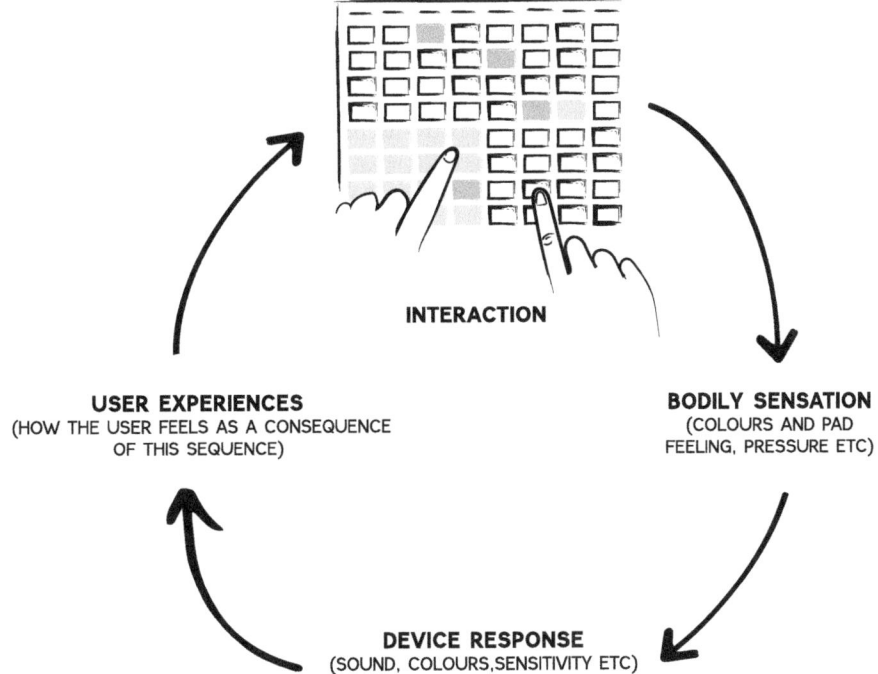

FIGURE 2.1  Illustrated by Emma Holdway

spend more time using it. The interaction becomes even more effective if it offers enough surprise elements and opportunities for exploration without overwhelming us. This loop can help us enter a flow-state, where we become completely immersed in our creative work. However, regarding technology, obstacles can permanently disrupt this interaction and leave us feeling insecure.

For example, even though Ableton Push as a controller can be highly interactive and enhance our creative confidence, we must first know how Ableton Live works. In my experience, Push can be a brilliant tool to use as the first interaction with a person who has never made music electronically, as it is effortless to approach. I have found that the 64-pad view of the instrument I use in workshops for children, people with disabilities, and those not comfortable with technology, is highly approachable. By setting up one instrument with headphones, people can easily slide their hands on the pads and interact with it in a comfortable way. This workshop approach has delivered a fantastic response and has become a meditative, empowering, and therapeutic tool. However, it is essential to note that this is a planned and guided session with an Ableton Live expert, not set up by the user.

So, even though modern design practices have made it easier for us to interact with electronic instruments and have improved our creative workflows, there are still many obstacles that users need to overcome before they can fully benefit from all the possibilities of these tools. This can be incredibly challenging for those who need more knowledge and experience to avoid getting stuck in basic setup and technical issues.

In Chapter 4 of this book, we will continue discussing how we can enhance the positive feedback loop using workflows. These techniques are designed to focus on the assistive and interactive features of the tools, and limit the chances of becoming overwhelmed or feeling insecure.

## Creative Confidence: Music Production Tools

Assistive music production tools aim to help us with obstacles in music-making, but surprisingly, they can still cause us insecurities. With assistive tools, we mean plugins or devices that, for example, are aimed to help us with a lack of knowledge in areas such as music theory, or can be used to speed up our creative workflow.

The issues with confidence can occur when the device removes the initial obstacle, but eventually, with or without the tool, the musician can have precisely the same problems with their workflow as before. They might still think that the progression is not good enough, the style does not fit the genre or the arrangement is not working. But are these fundamentally human problems even possible to solve with technology?

For example, let us say that you ask a person to draw a picture using only one pen. The creative task and the tool are simple, and there is no right or wrong outcome. It is the person who makes this straightforward task more complicated, with questions like how they should use the pen, what they should draw, and whether the drawing will be good enough.

In his TED talk, Sir Ken Robinson refers to a child who was in a drawing lesson and was asked by a teacher, "What are you drawing?" To which she answered, "I am drawing a picture of God". The teacher replied, "But nobody knows what God looks like", to which the child said, "They will in a minute".[8]

This little girl's story demonstrates our ability to have confidence in our perspective and vision, regardless of the instructions. The user's creativity defines the purpose and capability of the pen; the pen does not determine the outcome. Why would this be any different with technical tools and devices? A plugin to help someone who does not know music theory, only solves one particular problem. It gives a chance for the person to dip into musical structures that might not be available to them beforehand. But the plugin does not make the musician or their music *better*. The art they create might be fantastic with or without this plugin.

I had a conversation with my mother, Kristiina Turtonen, an artist with a background in educational theory, asking her about how education affects our creative confidence in arts. She answered:

> As educators, we give students the tools, techniques, theories and history to learn. But it is also our responsibility to teach them the courage to explore, to express their personality, perspective and vision. Without the first, the second cannot function and vice versa.[9]

If we expect that tools will give us results faster and easier, it might be just a way for us to shift the blame of our disappointment on to something that is far easier to understand, without dealing with the root of the issue. Where technological change has given us these incredible devices to help and assist us on our creative journey, we still need to remember that art and outcome are about human feeling and emotional connection to what we see and hear.

This topic will be further discussed and expanded in Chapters 3.5 and 3.6 of this book.

## Technology and Space: Confidence to Create

As modern music producers, our creative process can be very lonely. Even though projects can be entirely collaborative, the artistry is often divided into responsibility roles. For example, songwriters are in charge of bringing in the melody and lyrics of the song, the artist does the performance, and it is up to the producer to collect all this data and put it together as one. Then there are music producers like me, who produce their songs, doing all these roles simultaneously and alone.

When producers collaborate, the dynamic of the workflow is completely different. The process may involve piecing together a track step by step or improvising during sessions. However, even with electronic instruments, the responsibilities and dynamics of the workflow can be complex. The nature of the collaboration and its impact on creativity depends on various factors, including the space and ownership of equipment used. Despite the collaboration, the final song is often put together by one producer alone in the studio.

In music production education, the focus is on technical knowledge and outcomes. Creative methods are thought through imitation of existing works, therefore dismissing the systematical approach to develop students' personal creativity. Yet, in many music courses that do not focus on the technological aspects, creativity is often explored through musicianship and collaboration.

In the world of music, this type of isolated work process is unique. What makes music production different from a lonely fine artist working on a painting in solitude? In music production it is likely due to the complexity, knowledge, expertise, and immobility of the tools. To get quick, fast, and professional results with music production equipment, you need to know how they function, feel comfortable using them, and you need to be in a space that allows all the necessary equipment to be in a ready-to-use mode. This is so that the creative collaboration is not interrupted by technical issues. Anyone who has ever worked in audio knows that troubleshooting can be a mood-killer.

In an interview with social psychologist Dr Liz Dobson, we discussed confidence in audio and its relationship to the spaces formed by physical, social, and cultural environments.[10] They talked about how creating environments in which everyone can feel accepted, safe, and respected can make a person less self-conscious. This would enable them to concentrate on the music, flow (Csikszentmihalyi), creativity, and zone without the focus being on their presence in the space. Therefore, if rooms and communities we work in do not make us self-confident, we become self-conscious. Dr Dobson continues, explaining how presenting their research around confidence in music technology, one of the delegates commented that:

> There should not be a necessary pressure on the person in the space to develop their confidence, but have the pressure on the environment and space to change to enable that person to flourish.[11]

This outlook by Dr Dobson is something to which I relate. As a music producer, I love working in my home studio. I know every cable, connection, and if I ever work with others in this space, I always feel in control of the session. Therefore, every time I step foot inside a big studio, my initial reaction is defence and extreme self-consciousness, especially if I enter the studio as a music producer and engineer, instead of as musician.

This reaction can be caused by many factors, such as my inexperience in big studios, imposter syndrome, the fear of not being taken seriously and the history of the role of gender in the audio industry and music studios. The fact that I enter these spaces with my defence mechanisms activated does not mean I am not confident or unwilling to work in studios. But due to these significant factors that create the environment, I will feel self-conscious until I gain trust that I will be treated with respect and feel safe. Feeling entirely comfortable and relaxed in the studio will then maximise the full potential for my creativity and creative confidence.

The current lack of diverse gender representation in studios, audio communities, and other music industry spaces significantly impacts our insecurities in these industries. We currently live in a patriarchal society, wherein positions of power and/or social structures continue to be dominated by men.[12] This system also applies to music production spaces. A study by USC Annenberg Inclusion Initiative (2012–2020) found that 2.6% of the music producers are women, of which only a fraction are women of colour. This number might be surprising for many, but as a woman in this industry, it is not only a statistic but forcibly a big part of my career.[13]

The role of patriarchy is complex and nuanced, therefore, and in-depth discussion is beyond the scope of this book alone. But, a fundamental point resides in that people of all genders are affected by patriarchy, not just women. Just one example is the perception that people must conform to an unobtainable version of "norms", whether it be a type of masculinity or something else, which then places those who are unable or unwilling to meet these standards at risk.[14] Moreover, people identifying as a male can experience extreme pressure about knowledge and success, whereas a female-identifying producer might struggle with crippling imposter syndrome. The difference between these two scenarios is that where insecurities might make a man doubt their abilities, they stop a woman before they even explore what they could do – explaining the lack of women and gender minorities represented in these spaces.

Concerning gender, it is good to discuss this topic through the words masculine and feminine. It is important not to assume that all men are predominantly masculine and women are feminine. Instead, recognise that we all have a balance of both and varying perceptions of these gender norms. Audio spaces are generally masculine spaces and this can be for several reasons. One can certainly relate to how technology is associated with masculinity. Hansen (2022) states: "the practice of singing has been culturally coded feminine, and it has held a lower status than activities perceived to be more masculine, such as songwriting/composing, the mastery of an instrument, or the mastery of technology (for example, record producing)".[15][16]

Therefore, maybe the space affects our confidence not only because of our gender but how feminine we are or how feminine we wish to present ourselves. Often the more masculine you come across to others, the more you will get respect, acknowledgement, and appreciation. This is a common topic with people in the LGBTQ+ community and people who identify as a woman. For example, they might feel the need to dress more masculine in the audio spaces to be taken more seriously. This is something I personally identify with strongly. After cutting my hair short I felt a massive change in how people talked to me in studios. This allowed me to focus more on the audio and less on proving why I should be in that space, but at the same time I felt sad that I was having to hide my femininity and how I was embarrassed by it.

Similarly, people who are trans, non-binary, gender queer, gender nonconforming, or gender fluid will experience insecurities and obstacles in these spaces in their own way. In an interview with music producer Tobias Mark, they talked about audio spaces and how they affect their confidence as a gender queer person:

> If you are not out as a queer person in those spaces, and maybe you are not as comfortable in your queer identity as you would like, you might be defending yourself. And that gets magnified in these environments. Because when you are faced with vulnerability, you leave yourself open to criticism and being found out, especially if I was not ready to tell them. So what if someone finds out? Do they kick you out of that space? If I reveal who I am, I suddenly can not do the things I want to and need to leave. Music studios are traditionally masculine spaces and generally do not condone that kind of queer environment. So we are dealing with it our own way. And that is why maybe hiding your queer identity and who you are inside is for protection.[17]

The main argument you often hear about the lack of women and gender minorities in audio is "they are just not interested". In 2019, together with my colleague Emily Johnson, we started a company, Equalise Music Production, delivering music production courses for women and gender minorities. We proved this common statement wrong in the first set of courses, consistently selling out the spaces in our lessons immediately. Women and gender minorities are interested in audio, but the obstacle is learning it in an environment where they constantly feel self-conscious. If you struggle with confidence from systematic societal structures, it is hard to regain it if you do not have the safe space to encourage this emotional development.

The same way I have battled against preconceptions about being neurodivergent, I continue to fight against others' views about my gender in the audio industry. As much as I managed to flip the narrative on my self-identity, learning to believe that I could become a music producer, the journey has not been easy. Doing it without role models showing me the way and being consistently one of the only women in class, I still found my way into audio. And even though I do acknowledge that I am privileged in many ways as a white middle-class person, my journey has taken a lot of courage and confidence, and it still does.

But I do see a change happening, which suggests that the next generation of gender minorities will start their journeys with more confidence, guidance, and support. As much as gender should not matter, it still does and will continue to do so until the 2.6% has gained significantly more percentage and the "where are all the female producers" question is no longer asked. If we aim to change these spaces together for the better, it benefits every single person – the only consequence being more fantastic music.

The conversation about gender and how audio spaces affect all of us is an extensive subject and while it does form an essential part of this topic, it is far too complex to discuss in the detail that it deserves, in this book. To continue this discussion, look at the list of additional resources at the end of this chapter. There you can learn more about gender and audio spaces and find recommendations for audio communities aimed at women and gender minorities.

## CASE STUDY 2

To demonstrate the complicated relationships we have with technology and how it directly impacts us as music producers, let me introduce the second case study called Theo. We will use *The Quarter of Confidence* again in the demonstration (introduced in Chapter 1). They started their journey into music production as a teenager when they received a recording bundle as a Christmas gift to record guitar and vocals.

1. Heart: Self-identity, STEM identity, and imposter syndrome.
2. Brain: Self-defeating thoughts about their physical capabilities to learn.
3. Space: Peer support, influence of role models, and learning environment.
4. Action: Interface interaction, exploration, and positive feedback loop.

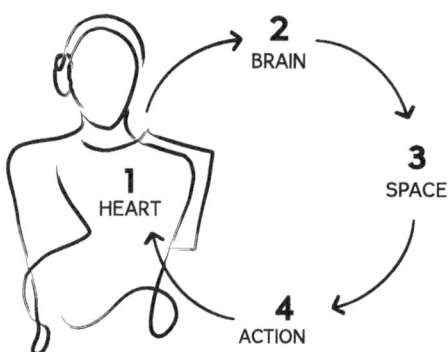

FIGURE 2.2 Illustrated by Emma Holdway

- **Heart** – This is where Theo has their identity as a learner and musician, and their first response to technology. Music studios have always seemed a little bit scary to Theo, due to the complexity of the equipment and the fact that they have not often seen role models they could relate to in these spaces. That is why, for example, enrolling on a university course to learn about production feels like it is outside of their comfort zone. Recording and production seem like fascinating practices, but at this point, Theo does not believe they could be a professional in the audio industry (self-efficacy).

  Theo comes from a family where no one is artistically inclined. Their parents have supported music as a hobby, but Theo still feels like music production will be too difficult to make a stable career. It would be better to keep it as a hobby and find a more stable job to do full time, especially as Theo has never been too good at STEM subjects at school and becoming professional would require more technically inclined talent. This thinking is based on identity, particularly STEM identity, imposter syndrome, and inherited values.

Imposter syndrome is the inability to believe you deserve success or that you belong into a space, feeling everyone else is more equipped and more worthy of being there. Whereas inherited values means opinions, through processes and beliefs inhered from family and the community we grew up with, moulding the way we perceive the world around us and our role in it.

- **Brain** – As much as Theo is interested in music production, they are still convinced that their brain is not wired to understand complex technical and audio concepts. This way of thinking has prevented Theo from pursuing an interest in audio before and now can make them feel insecure in the learning process, leading them to believe they do not have what it takes to be at the same knowledge level as professional music producers.

   With more self-awareness, knowledge of different learning methods, and understanding of cognitive processes, Theo could find the best way to learn these concepts. For example, maybe the way the school system taught physics did not support how Theo could understand it, making it feel like it was their fault and that they were incapable of learning. Therefore, Theo would need to find more information on their unique way of learning and focus on the good qualities of their learning process.

- **Space** – Theo has been part of an audio and music production social media group for many years. They believed that this group could give them ideas and inspiration for learning. The group has been helpful in many ways, but it has also made Theo compare their skill sets, learning process, and equipment to others. They also do not feel comfortable asking questions in the group, as Theo has witnessed some insulting comments on other group members' posts. In the group, Theo does not see many other people they could relate to, which makes them feel like an outsider, imposter, and like they are not part of the community.

   In this scenario, Theo has not been introduced to the audio industry through people with whom they would feel entirely comfortable and the group experience has become more about being self-conscious around others in the community and less about music production itself. This has influenced Theo's decision to start practising production at home and explains why they feel uncertain about applying to a university course. They wonder if in the classroom they would feel like as much an outsider as they

> do online. What if there is no one to relate to? What if everyone else knows more about music production already? And what if they do not have enough equipment to produce at the same level as others?
> 
> - **Action** – In Theo's story, there are plenty of obstacles that could prevent them from pursuing music production as their hobby or considering it as a career. Through the encouragement Theo received from their parents via the new interface, they have an opportunity to explore, enjoy, and create in their own time. Through this practice, they will find confidence in their skills and increase their passion for audio and music production.
> 
>   The interface and software inspired Theo to explore free music-making apps, allowing them to develop some of their guitar-based compositions into complete tracks. Despite the overwhelming number of devices to explore, Theo has found enjoyment in these simple tools that are either free or that came with the recording bundle.
> 
>   A couple of years after receiving this Christmas gift, Theo found a community of people, friends with similar passions, and together with them, applied to a university degree to study music production. All of the same insecurities still exist in Theo's everyday practice, but with the support of these new friends and feeling more secure with their equipment, Theo has found the confidence to tackle these negative feelings.

This case study demonstrates a story, possibly the most common one that I have encountered in my career. No story is similar and we all have our own obstacles on our journey, whether it is at the beginning or when we are professionals in our careers. But what connects us all are these complex patterns of life, perception, and trying to understand what is best for us and what we really want.

The following chapters will look deeply into how confidence, creativity, and technology manifest in our journeys, through our most common insecurities. The chapter "Insecurity Corner" aims to help you understand and navigate your mental game between the signals, while providing reflection that we can all relate to.

## Notes

1. Mark Ronson. Mark Ronson – Uptown Funk (Official Video) Ft. Bruno Mars, 2014.
2. Marie Tricaud. (September) 2021. Interviewed by Liina Turtonen.
3. James Tuck. (January) 2024. Interviewed by Liina Turtonen.
4. The Interaction Design Foundation. n.d. "What is 21st century design?". Accessed 15 January 2024. www.interaction-design.org/literature/topics/21st-century-design.
5. The Interaction Design Foundation. n.d. "What is User Experience (UX) Design?". Accessed 15 January 2024. www.interaction-design.org/literature/topics/ux-design.
6. The Interaction Design Foundation. n.d. "What is design thinking?". Accessed 15 January 2024. www.interaction-design.org/literature/topics/design-thinking.
7. The Interaction Design Foundation. n.d. "What is Human-Computer Interaction (HCI)?". Accessed 15 January 2024. www.interaction-design.org/literature/topics/human-computer-interaction.
8. Sir Ken Robinson. 2006. *Do Schools Kill Creativity?*, TED. https://www.youtube.com/watch?v=iG9CE55wbtY
9. Kristiina Turtonen. (September) 2021. Interviewed by Liina Turtonen.
10. Dr Liz Dobson. (September) 2021. Interviewed by Liina Turtonen.
11. Dr Liz Dobson. (September) 2021. Interviewed by Liina Turtonen.
12. "Patriarchy – An overview | ScienceDirect Topics". Accessed 12 November 2023. www.sciencedirect.com/topics/social-sciences/patriarchy.
13. This paragraph is written in collaboration with Sophie Russell, November 2023.
14. This paragraph is written in collaboration with Sophie Russell, November 2023.
15. Kai Arne Hansen. 2022. "Staging a 'real' masculinity in a 'fake' world: Creativity, (in)authenticity, and the gendering of musical labour".*Cultural Studies*, 36(5): 713–731. https://doi.org/10.1080/09502386.2021.2011932.
16. This paragraph is written in collaboration with Sophie Russell, November 2023.
17. Tobias Mark. (April) 2022. Interviewed by Liina Turtonen.

## Additional Resources

- Barker, M.-J. and J. Scheele. 2016. *Queer: A Graphic History*. Icon Books; illustrated edition.
- Bjørn, K. 2017. *Push Turn Move*. Bjooks.

- Bjørn, K. and C. Meyer. 2018. *PATCH & TWEAK: Exploring Modular Synthesis*. Bjooks.
- Csikszentmihalyi, M. 1996. *Flow: The Psychology of Optimal Experience*. Harper & Row.
- DeSantis, D. 2015. *Making Music: 74 Creative Strategies for Electronic Music Producers*. Ableton.
- Fine, C. 2011. *Delusions of Gender: How Our Minds, Society, and Neurosexism Create Difference*. W. W. Norton & Company; reprint edition.
- Iantaffi, A. and M.-J. Barker. 2017. *How to Understand Your Gender: A Practical Guide for Exploring Who You Are*. Jessica Kingsley Publishers.
- Jordan, P. W. 2002. *Designing Pleasurable Products: An Introduction to the New Human Factors*. Taylor & Francis.
- Jordan, B. 2021. "Why aren't there more female music producers?" YouTube, https://youtu.be/2Ipb81z46kI.
- Kelley, D. and T. Kelley. 2015. *Creative Confidence: Unleashing the Creative Potential Within Us All*. Crown Business.
- LNA Does Audio Stuff, Music Production for Women. 2021. "IN CTRL: A journey into music production". YouTube: https://youtu.be/ekuXlf2OVj8?si=Z9Eb58K-13fHGshq.
- Marie, K. 2022. *Conversations with Women in Music Production: The Interviews*. Backbeat.
- Taylor, J. 2012. *Playing it Queer: Popular Music, Identity and Queer World-Making*. Peter Lang.
- Todd, M. 2016. *Straight Jacket*. Bantam Press.
- Wolfe, P. 2020. *Women in the Studio: Creativity, Control and Gender in Popular Music Sound Production*. Routledge.
- Women and Equalities Committee. Misogyny in Music. House of Commons. Second Report of Session 2023–24 HC 129, https://committees.parliament.uk/publications/43084/documents/214478/default/
- Zwaan, J. "Josephine Zwaan". https://uva.academia.edu/JosephineZwaan.

# Insecurity Corner 3

## Introduction

### Keywords

- defense mechanisms
- conscious mind
- unconscious mind

### Excuses, Reasons, and Insecurities

Sometimes we can feel stuck with our life. We can lose direction and are unable to find a balance between our desires, dreams, and the responsibilities and realities of living. Or we are simply surviving day by day without the capability to dedicate time for ourselves. That is why creativity is often one of the things we find easiest to cut out from our schedules and leave for another day. We find it challenging to prioritise artistic practice, and when we do, we struggle to navigate our goals and passions within the world surrounding us. When we constantly put off pursuing our goals and desires, it becomes increasingly difficult to motivate ourselves to start taking action. This is because our brain has a tendency to focus on the negative evidence of us not doing it, rather than the positive evidence of us actually taking action. Over time, this negative focus can lead to feelings of insecurity and a lack of confidence, which in turn makes it easier to come up with excuses for not pursuing our full potential.

The upcoming chapters will delve deeper into why we sometimes feel insecure and how it can manifest in our everyday lives. The conversation is directed by our psychological defence mechanisms, for which we often use another name: excuse. Psychology has identified several defence mechanisms that can show up in different ways. For example, denial is a

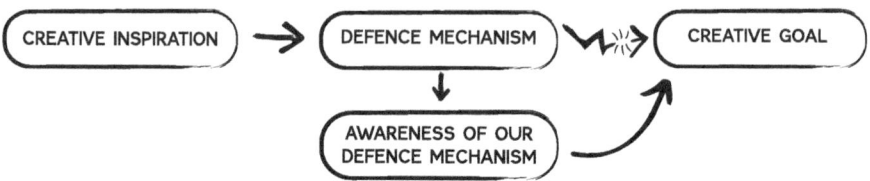

FIGURE 3.1 Illustrated by Emma Holdway

defence mechanism, where a person might find technology so overwhelming that they refuse to try or experience it in the first place.[1] These unconscious habits can limit our potential and prevent us from seeing all the possibilities. Identifying these harmful habits can help us balance our personal and creative lives better.

We use defences to protect ourselves from things that might make us uncomfortable or scared. For example, one common defence might be "I do not have a techy brain," although you have been practising production for years. A person declaring a strong defensive sentence like this to themselves or others is a way to ensure no one is expecting from them more than their confidence is allowing.

The word *excuse* can seem a bit harsh, but it only feels negative due to the *reasons* and *justifications* we give it. For example, if we truly believe that we have no capabilities to learn anything technical, then we have given ourselves a reason for not pursuing these skills. For us this reason might feel completely justified, and therefore it can be challenging to accept that the only person holding you down from pursuing technical knowledge is yourself. But are we capable of seeing when our reasons have become excuses? We often think that the reasons we give for our behaviour are the truth. This is called *rationalisation*, and it means that we tend to believe human actions are consciously motivated, whereas, in psychology, this has been discovered to be mostly untrue.[2]

The psychoanalytic theory of personality by Sigmund Freud defines how we are impacted by unconscious thinking; having feelings, thoughts, urges, and memories outside of conscious awareness.[3] In simple terms, this means we are not always aware of why we think, feel, and act a certain way, which can then force us to rationalise and explain our difficult emotions with defence mechanisms.

In the previous two chapters, we discussed how we can switch up the narrative of our self-identity and gain confidence through awareness of ourselves. Now is the time to start recognising our defence mechanisms and stop giving them the power to stop us from achieving our goals. It is the biggest challenge, but the most rewarding thing we can do for ourselves.

## CASE STUDY 3

Robin has been doing music production for over 15 years. He knows software, synthesisers, and plugins like the back of his hand, but he has never released any of his own music. Robin has a partner and three children, and he also works as a full-time music teacher. Music has been his dream career for as long as he can remember, but it feels like an impossible and unrealistic dream because of time and finances.

Because Robin has bills to pay and bears the responsibility of supporting a family, a teacher position brings comfort and safety to everyday life. This is a reason not to change job in pursuit of a dream, but it can become a defence mechanism because of the conscious and unconscious fears a change could bring. Otherwise, Robin might feel insecure in his own ability to achieve what he wants or succeed in being a good enough producer to make it a sustainable career. This is where Robin's brain might find plenty of evidence from the past to support why change is not a good idea, increasing his fears and making an alternative future seem harder.

Similarly, all of the factors mentioned above contribute to Robin's confidence as a music producer. He hardly has any time to practise, and when he does, it becomes hard to finish a track as Robin feels he is out of touch with the technology, maybe others know more than him, or perhaps the music industry will not accept his homemade songs. That is why he does not practise as often as he wishes and struggles making time for his creativity.

When a person feels stuck in their life situation they can struggle to see any other path, choice, or option for their future. In this scenario, Robin's main issue is not the career shift but the journey he would need to undertake to achieve a feeling of accomplishment in his dreams and desires. Whether Robin will always need to keep a job as a teacher is secondary. The focus should be on his journey to discover his unconscious thoughts about himself and how those translate to defences about his career and life direction.

By gaining self-awareness, Robin might find balance in being a teacher, whilst being confident in making music and releasing it in his free time. Or maybe one day, after practising both confidence and production skills, he will find the courage, time, and support to pursue his dream career. Either way, there is no right or wrong way to feel the feeling of fulfilment. Success is already finding confidence in your creative visions.

## What Is in the Insecurity Corner?

In the past two chapters, you have been introduced to some of the concepts of creativity, confidence, and technology. The Insecurity Corner will continue these discussions, diving deeper into why we might feel insecure as music producers, music-makers, artists, and participants in the music industry.

The chapter titles are formed from excuses and statements I have heard the most during my professional career. The Insecurity Corner is divided into two different sections. The first sections (Chapters 3.1 to 3.8) are the most common insecurities we have all gone through at one point in our journey. Most of these will be something that beginners, or people who might not even be in music production yet, might relate to. Nevertheless, these chapters withhold some of the most common insecurities that are hard not to connect with.

The second group of chapters (3.9 to 3.12) looks at some of our deeper approaches to the elemental insecurities we might have in creativity, which help us understand our insecurities and defences on a more psychological level. These include perfectionism, jealousy, envy, validation, and fear. In most of these chapters, there will be exercises to try out, to help find identify your defences and insecurities and how they affect your creative practice.

## Notes

1 Robert Hogan, John Johnson, and Stephen R. Briggs, eds. 1997. *Handbook of Personality Psychology*. Academic Press.
2 Jeremy E. Sherman. (19 November) 2015. "What's the difference between rationality and rationalizing?" *Psychology Today United Kingdom*. Accessed 15 January 2024. www.psychologytoday.com/gb/articles/what-s-the-difference-between-rationality-and-rationalizing.
3 Kendra Cherry. (19 June) 2020. "What is the unconscious (and why is it like an iceberg)?" *Verywell Mind*. Accessed 15 January 2024. www.verywellmind.com/what-is-the-unconscious-and-why-is-it-like-an-iceberg-2795901.

# I Do Not Have a Techy Brain

# 3.1

## Keywords

- techy brain
- STEM identity
- social identity threat

## The Power of Self-identity

We are going to start with the most common excuse: the *techy brain*. As an educator, these words are very familiar to me, and I do not think there has been a course where I have not heard at least one person say this at the beginning of the class. For example, the *techy brain* concept is something I remember hearing my mum say many times when learning to send emails. Similarly, when I mention being a music producer to a new person, and they respond with, "that sounds clever, I do not have the *techy brain* for that type of stuff".

Many people genuinely believe this about themselves, but it's possible that this is actually a defence mechanism that arises when they feel uncomfortable in a situation. As humans, we tend to rely on habits and defenses to feel secure in our identity. Depending on the case, this type of behaviour can be part of an experience called *social identity threat*.[1] People will try to make themselves feel comfortable in a situation by undermining their intellectual capabilities. If a person lacks confidence in their abilities or knowledge regarding technology, it is common for them to feel nervous or uncomfortable when faced with a conversation about it. They may respond by downplaying their skills, making it seem like they are not capable of contributing to the discussion. This can be a way to avoid pressure or expectations from others.

Similarly, a *stereotype threat* is a psychological phenomenon where we start to believe in the stereotypes we have been told since birth.[2] One common

DOI: 10.4324/9781003194484-5

stereotype is that women are less capable of technological skills than men. It is a stereotype formed by the history of gender roles in society, representation, and patriarchal social structures. Therefore, as the idea of stereotype threat describes, women might believe this argument when faced with technology. They undermine their skills by saying sentences such as "I just do not have a techy brain". The stereotype and identity threats can make it very difficult for women to start practising technology. If they do, they might constantly feel the need to prove to themselves and others that they are capable of achieving something, which can lead to inner battles. This phenomenon is commonly known as *imposter syndrome*.

When interviewing industry professionals for the last chapter of this book, we started a fascinating conversation about *techy brain* with producer and sound designer Kelly Buckley. After discussing her career and expertise in the industry, she told me aside that she does not think she is a technological person. This sparked a conversation where I mentioned that, in my perspective, she is a technologically inclined person, as she does audio as her job and uses all of these technological tools in her art. This is one of the reasons I asked her to be interviewed in this book. For this conversation, Kelly answered:

> So even though I am hearing this, I still do not think I am a techy person. It is so deeply ingrained. Even though I am going, "yes, you are right," and I can see why I feel like this. But when I take it back to myself, I question it. Although, in the end, I must be a technological person, even if I do not always feel like it. I have spent years learning audio and spent hundreds – thousands of pounds even – on the lessons and equipment. And I would still not easily call myself a producer. But I am learning to be better at that because I have to. I need to be confident to say what I am, especially when I discuss with clients about doing their soundtrack.[3]

The sentence "I just do not have a *techy brain*" is so powerful that it can make someone write off their own abilities before they have even explored the technological tools available to them. However, this concept is not rooted in reality. There is no such thing as a "technological brain". Your ability to learn STEM (Science, Technology, Engineering and Mathematics) subjects depends on several different neurological factors in your brain, making you process information in different ways.[4] The most significant part of your capacity to learn these topics is about tutoring and what learning resources are available to you and how you learn best.

For example, as a person with dyscalculia and dyslexia, I found mathematics in school almost impossible to learn, as my brain cannot process numbers without any visualisation. However, later in my Master's degree, I was asked to code a plugin where I needed to use mathematical equations. The only resources I found compelling in my learning were YouTube videos with titles like "Mathematics for Dummies". These tutorials were relatable and fun, and I did not feel insecure while learning from them. They used plenty of graphics and humour to explain the concepts. Therefore, with knowledge from my university lectures, I gained the confidence to change my perspective on my identity as a STEM learner. This resulted in learning the maths needed to code an LFO (low-frequency oscillator) using JS programming language. My struggle with the basics of maths from school does not make me bad at mathematics, and it does not mean I could not identify as a mathematical person. I just needed to understand how my brain functions to learn the requested skill. The same applies to technology and all the sciences.

From a young age, we take the role we are given without questioning. We can be more than that. We start to believe in these roles and form our identities around them. As humans, we are far more complicated than the social cliques in schools or the positions we take in our families and society. But as much as we learn to believe something about ourselves, we frequently judge others similarly. Often unconsciously. It is important that we avoid categorising people based on their external features, how they talk or act. At the same time, we should allow ourselves to expand our own beliefs about what we are capable of and what we could become.

If you have ever caught yourself thinking that you cannot learn something, think again. Maybe your difficulties are more to do with the tutor or learning material and not you. It would be an absolute shame for anyone not to discover the potential that technology offers us as individuals. We all have a *techy brain*, and if your brain is so powerful that it can make you conscious, speak, and remember, then do not undervalue its capabilities. Give it a chance to discover what it can do.

---

### HOW TO GET OVER TECHY BRAIN INSECURITIES

1. **Say and write down your new identity**
It can be a defence mechanism not to call yourself by the title you deserve. Learn to put into words what you wish to be and who you truthfully are:

- I am a music producer.
- I am an audio engineer.
- I am a technical person.

It might be that you feel like an imposter (experiencing imposter syndrome) the first time saying this, or you might even feel shame and embarrassment. Try to do this despite feeling slightly uncomfortable; remember, this takes practice. The more you say it, the more you start to believe in it.

2. **Focus on what you know, not on what you don't already know**
The first step towards changing your relationship with technology is to approach it at your own pace. There are all sorts of tools, software, instruments, techniques, and fun gadgets to create sound within music production. This can be very overwhelming and put people off before they even start to learn about any of them. Therefore, start with something you already know and feel comfortable with. Maybe your phone or a free app on your computer? Focus on the enjoyment of making music and the sounds that come out of this tool.

Once you feel comfortable using this music-making device, try exploring other similar instruments as well. Keep learning, enjoying, and researching all the tools that you use. Before you know it, you'll be creating music with devices that you might have only dreamed of using before. This is how you can change the narrative on your self-identity by taking the first steps and continuing to be curious.

The more knowledge you obtain, the more you realise just how much more there is to learn. Therefore, whatever stage of learning you might be at, remember these facts about the audio industry:

- The *techy brain* does not exist, and everyone can learn to use technology.
- No one knows everything, and everyone starts from nothing.
- If the song is fantastic, the tools that it was made with do not affect its value.
- If you make music with *any* electronic equipment, you can call yourself a music producer.
- Do not let anyone tell you what you can and cannot do.

3. **Be kind to yourself and have patience with both your learning and outcome**
In the beginning, learning technology can be frustrating, and there are plenty of moments where you might blame yourself for equipment

not working. I am happy to tell you that most of the time, it is not your fault.

One of the first things each audio course should teach is the troubleshooting list for the equipment. The longer you work with these tools, the more you will memorise this list and go through it every time there is no sound, or a button that does not work the way you think it should. In those moments, ask for help and be kind to yourself. Even if the solution to the problem has a straightforward answer, remember we have all gone through it. I still remember the first time I recorded a live concert and forgot to turn the performer's microphone on. It still hurts to remember, but I moved on, and I have never forgotten to double-check power buttons again.

Similarly, in the beginning, your songs and beats might not sound anything like ones on the radio. Again, be kind to yourself and remember you will get to those results by exploring and being curious. The people who made those songs were in your position at one point. That is why the best thing to do is to concentrate on sounds you like and find beauty in the enjoyment you experience when creating. The technical know-how will come with time.

When you feel like an outsider or when it is hard to stay confident in your practice, seek peer support, as sometimes we need a reminder that we are not alone in our journey. It is important to find communities and groups where you feel safe learning. Remember to be selective with these groups as well, so that you find the ones that genuinely only lift you and do not make you feel more insecure.

That is why it is good to remember:

- Mistakes and troubleshooting are everyday practices. You will learn to deal with them with time, and most likely, they are not even your fault.
- Turn on power buttons.
- Every single professional in audio has made all the same mistakes you are making now.
- There is no hurry to get to your goals, and remember, you will never stop learning.
- Seeking peer support is not a weakness but an opportunity. Look for people that encourage you to go forward, who keep you accountable in your journey and who inspire you along the way.

## Notes

1 Naomi Ondrasek and Lisa Flook. 2021. "How to help all students feel safe to be themselves". *Greater Good*. Accessed 15 October 2021. https://greatergood.berkeley.edu/article/item/how_to_help_all_students_feel_safe_to_be_themselves.
2 "Stereotype threat". n.d. *National Institution of Health*. Accessed 15 January 2024. https://diversity.nih.gov/sociocultural-factors/stereotype-threat.
3 Kelly Buckley. (August) 2022. Interviewed by Liina Turtonen.
4 Linda Marsa. (14 January) 2014. "Why some of us are better at math than others". *Discover Magazine*.

## Additional Resources

- Cheryan, S., V. C. Plaut, P. G. Davies, and C. M. Steele. 2009. "Ambient belonging: How stereotypical cues impact gender participation in computer science". *Journal of Personality and Social Psychology*, 97(6): 1045–1060.
- Cultivating STEM Identity Panel Recording. 2022. Accessed January 2024. https://youtu.be/A6WwXURVxiE.
- Fine, C. 2010. *Delusions of Gender: How Our Minds, Society, and Neurosexism Create Difference*. W. W. Norton & Company.
- Shetterly, M. L. 2016. *Hidden Figures: The Story of the African-American Women Who Helped Win the Space Race*. William Morrow.
- Singer, A., G. Montgomery, and S. Schmoll. 2020. "How to foster the formation of STEM identity: Studying diversity in an authentic learning environment". *International Journal of STEM Education*, 7, Article number 57.

# 3.2 Audio and Music Production Is Difficult

## Keywords

- gatekeeping
- elitism

For some, music production and audio engineering can seem extremely difficult subjects to learn and practise professionally. In this chapter, we will look into the reasons behind this perception, which lies in wide range of reasons. We have previously discussed the impact of audio spaces on creativity and confidence, as well as STEM identity, both of which are relevant to this topic. Additionally, we will discuss how our society tends to value more *difficult* pursuits, often placing greater importance on engineering rather than art. We will delve further into these issues to gain a better understanding of the challenges faced by those pursuing music production and audio engineering.

## What in Audio and Music Spaces Makes it all Seem so Difficult?

When entering the audio environments, often we feel like we do not know enough, and that everybody else knows more. The topics can seem very overwhelming and complicated, and you might not know where to start. It might look like others have had a head start, and our brain is the only one that cannot keep up. You might start to question if you have the capacity and time to learn it all. And sometimes these feelings might continue all the way to the professional life.

DOI: 10.4324/9781003194484-6

Technology has become an essential tool in the creative process, and the perceived complexity of devices in music production and audio engineering can make mastering these subjects a source of pride. However, this pride can sometimes lead to competitive boasting in social situations. While it is natural to take pride in one's accomplishments, boasting can also stem from feelings of insecurity.[1] This is where the audio and music spaces can be prone to competition, social hierarchy, gatekeeping, and types of elitism. Therefore, the different emotional triggers can lead to students avoiding asking questions in fear of sounding stupid. For many, this atmosphere in classrooms, studio spaces, and within online communities can make music production seem more complicated than it truly is.

The Dunning–Kruger effect is a psychological theory about knowledge and confidence.[2] It explains why the knowledge levels in music and audio spaces can often feel competitive (refer to Figure 3.2). According to this theory, the faster people learn a new skill, the faster they will gain great confidence. The more you keep on learning, the more you start to understand how much more there is still to learn. And this is where you can start to doubt yourself and your skills, especially compared to others. But if you keep on practising and learning, you will gain confidence in your knowledge and in your learning process. In this last phase of learning, you might

FIGURE 3.2 Illustrated by Emma Holdway

reach the same level of confidence you had at the beginning, but with more sustainable results.

Now, if we compare the Dunning–Kruger effect to the music and audio industries, we can find similarities. We often presume technology as a very difficult subject to learn. Therefore, when a person finds success in learning how something like a compressor or modular synthesiser works, they might feel great confidence in their new knowledge. Now consider a classroom, workplace, or an online forum full of people feeling this extreme confidence in their skills. Naturally, this would make any space feel competitive or unwelcoming.

Knowledge does not make a person a *better* producer. It makes them a more *developed* producer. Anyone can learn to record, play instruments, use MIDI, and arrange tracks. The tools for music production are made for regular people to use, and you do not need an engineering degree to understand how a beat-making phone app works. The in-depth knowledge will come with time and after learning what areas you wish to focus on.

## Support, Personal Responsibility, and Determination in Learning

Learning new things is always about going outside of your comfort zone, and this experience can be sometimes extremely scary. But it is good to remember that the feeling of anxiety that we experience in these moments is normal and is nothing that you cannot overcome. This is what Anxiety Canada says about the topic:

> The process of facing fears is called Exposure. Exposure involves gradually and repeatedly going into feared situations until you feel less anxious. Exposure is not dangerous and will not make the fear worse. And after a while, your anxiety will naturally lessen.[3]

Negative emotions are part of life and a natural self-defence mechanism when we are reaching out to new dimensions psychologically and physically. The key is to accept these feelings, notice them, realise that they do not define you, and find a way to complete the task you started. But as much as we can accept negative emotions and feel ready to face our anxieties, we all need support to lift us. This can be a family member or a community, but it is not their responsibility to carry us fully. Support means someone who listens and reminds you of your full capacity when you forget it. But in the end it is your responsibility to take care of yourself and your future.

## Author's Experience

In the beginning, one of the biggest motivations for me to learn music production was the need to prove I could do it as well as (or even better than) the boys in my university class. I feel sad that such a negative motivator was behind my early learning, but it made me forget my fears and focus on becoming as good as possible at my craft.

Most of the time, I put on a brave face and made sure no one would see any weakness. I did not want to be the girl who struggled and asked stupid questions; I wanted to be treated the same as everyone else in the class. But behind closed doors, I truly felt my insecurities. Most of the time, I did not even notice how overwhelmed I was, but my husband did. He was the support network I needed when I felt the most insecure about my skills, studio sessions, or tracks. This support was vital in giving me the strength to keep up with learning and not give up. Because when we do not have the power to believe in our capabilities, we need someone next to us who will.

Later, going into my Master's degree in music production, I felt confident about my abilities. However, I quickly realised how behind I was with my audio theory and sound engineering knowledge, as my undergraduate degree had left significant gaps in my experience. Once again, my self-confidence and self-esteem were low.

As mentioned in the previous chapter, we needed to code a plugin in Reaper using JS language for this university course. As someone who had never done advanced mathematics or coding, I felt full of anxiety and panic. How on earth could they expect someone like me to learn how to do this? I ran crying to my tutor and started listing all the excuses I had gotten used to relying on; I am neurodiverse, and coding is way too tricky for my stupid brain.

When you have grown up with a diagnosed learning disability, you have gotten used to a certain level of forgiveness in education. You usually get extra time and some leverage with the mistakes. It becomes a very convenient reason to use when you feel insecure. My course leader listened to my very dramatic explanation of my weaknesses with patience and then said: "That is okay, so you just need to study harder".

His blunt response changed my perspective forever.

He was there to assist me in my journey, give all the resources and help needed, but it was not his job to make sure I learned. The responsibility was entirely mine. I was afraid, scared, insecure, a little bit angry, and I wanted to escape, but with the support at home and from my lecturer, I did learn to code. It took all my evenings and weekends at the library, and it was a painful process, but with a determination not to quit, passion towards audio, and the need to show to myself that my brain is no less capable than others', I did finish it. It was not a great or fully working plugin, but it was a start.

## Exercises for Learning a New Skill

1. Change your learning style.
   Just because you didn't understand something taught in a particular manner, does not mean you can't learn it altogether. To improve your learning experience, it is important to identify the type of learner you are. Consider how you can understand complicated topics better and in what format you prefer the lessons to be delivered. Here are some alternative learning options:

   - See if there is an app for learning the topic in mind. For example, there are several different apps for learning music theory. These apps are great for when you have a moment to spare. They are accessible and less intimidating than organised courses.
   - YouTube is always a fantastic place for learning anything. Remember to use the time to find the most suitable teachers for you, and do not give up if the first few still make you feel intimidated or overwhelmed. There is someone for everyone.
   - You can test your learning style. In the past decades, many psychologists have developed tests to see which way you learn the best. Even though it is excellent to know what is the best what you will learn, I would take these tests with caution. Some can be dated and often limiting for the learner if used more as a fact than a guideline. Below, you can find a link to an excellent article by Mind Tools that explains what tests are available, and they share great tips on finding the best learning style for you.

2. Let your passion for the topic guide you and be curious.
   Curiosity is the best starting point for learning a new thing, even if it seems scary or difficult initially. If there is a particular topic you are very interested in, find your ways to research it. Maybe talk to people who

might know about it, have conversations and ask questions. You can also learn a lot by joining communities, events and seminars while meeting like-minded people. Be cautious about surrounding yourself with only people who spark your curiosity, rather than put it down.

3. Repetition, time and patience.
Remember to dedicate time to the topics you find the hardest and find patterns of repetition to go over these topics. Remember, there is no hurry to learn anything, and you are not in competition with anyone else. For example, it took me five years and two degrees to fully understand what compressors do. I always felt like everyone else knew more than me and faster, but the fact is that we are all on our timelines, and learning is an endless marathon and not a race.

## Notes

1 R. B. Joelson and D. S. W. Joelson. (March) 2018. "Pride or boasting". *Psychology Today*. Accessed 15 January 2024. www.psychologytoday.com/gb/blog/moments-that-matter/201806/pride-or-boasting.
2 J. Kruger and D. Dunning. 1999. "Unskilled and unaware of it: How difficulties in recognizing one's own incompetence lead to inflated self-assessments". *Journal of Personality and Social Psychology*, 77(6): 1121–1134.
3 Anxiety Canada. n.d. "Facing your fears: Exposure". Accessed 15 January 2024. https://www.anxietycanada.com/sites/default/files/FacingFears_Exposure.pdf.

## Additional Resources

- DeSantis, D. 2015. *Making Music: 74 Creative Strategies for Electronic Music Producers*. Ableton.
- Mind Tools Content Team. Learning Styles. Mind Tools. Accessed January 2024. www.mindtools.com/addwv9h/learning-styles.

# I Am Too Different to Fit In

# 3.3

## Keyword

- authenticity

## What Is to Be Yourself?

As music-makers in the modern era, it is common to feel conflicted about our identities. On the one hand, we aim to develop a unique sound and voice that sets us apart from others. On the other hand, the music industry often pressures us to conform to certain standards in order to achieve success. Whether you are signed to a record company or a self-publishing artist, you must be ready to become a brand. In popular music, rock, or alternative genres, you need to be familiar enough so that people feel comfortable approaching you, but also so unique that what you create is seen as new and fresh. And in the midst of this, you are told: just be yourself, and that will be enough.

But how can you be your true yourself and what does it mean? In social media, we follow the artists we love. We relate to some of their characteristics and wish we could be like them. But how do we recognise the difference between what we think we should be and who we want to be? We often create unrealistic expectations based on what we believe will be popular and what will make us feel accepted, therefore fearing to show our true personality in our communities and on social media.

Being thoroughly ourselves is difficult already as a child. We start to monitor our emotions and adapt them to behaviours that seek acceptance from our guardians and the people around us. When we feel big emotions that might be confrontational to our parents, such as anger, we might see how it affects them, and the next time we will adapt to what makes them more comfortable.

The video "The True and the False Self" explains where our need to belong comes from: "Because our caregivers were preoccupied or fragile, we had to be preternaturally attuned to their demands, sensing that we had to comply (as a child) in order to be loved and tolerated; we had to be false before we had the chance to be properly alive".[1] This is why we might feel a duty to do the tasks we are asked for or think are expected from us, but it can make us feel uncreative and unoriginal, as we are scared to break from the mould. If we do something outside of the box, maybe it will make others see our authentic selves? Perhaps they do not like the real us?

Therefore, to find your authentic voice as an artist, we need to find ways to stop constantly trying to please the people around us. We must stop seeking love and acceptance for what we create. Because, if we are scared to be difficult and untempered about how our emotions might make others feel, how can we make art that is not filtered by the *space* around us?

As humans, we are very intuitive to the fact that someone is not being authentic. The same goes for music. If we cannot sense the real emotions behind the song's performance, expression, and lyrical content, we find it hard to connect with it. This does not always mean that the artist who performs the song is the writer, but to create a performance that has an impact, they might need to find a way to connect with the meaning and interpret its message through their feelings.

It is also important to mention that finding a way to be authentically yourself and the popularity of your art are two separate topics. Authenticity does not always mean that you cannot make popular music that imitates what others also make and what gets into the TOP10 playlists. It also does not mean you must avoid popular sounds and separate from the mainstream music industry. Being yourself means acknowledging what is healthy for you, what you enjoy, and that you are working towards the goals you truly desire. It is pure honesty with yourself, where you navigate your career and choices not to please others, but to impress only yourself.

In an interview with music producer and performer Rachel K. Collier, she was asked what advice she would give to anyone seeking a career in music, music production, audio or social media, for which she answered:

> Just be yourself. Do not live for trends or anything that you think will just get you famous quick. As much as fame can quickly come along, it can disappear just as quick. If you craft and create something unique and special, you will build an authentic audience and real fans that will stick with you for much longer.

If I had just made straight-up EDM, it might have been more straightforward to promote my music – for example, on places like Beatport or Tracksource. But I was sitting with my patrons the other day and they were literally saying that they love me because my music is not like anyone else's. This was what I need hear from time to time. It reassures me that I am doing something right.

So I would always say be true to yourself. If you want any chance to sustain your career, you need to love it, and be able to do it day in day out and devote your life to it. If a tiny part of you does not enjoy what you're making or creating, you will not be able to carry on. When you love it, you can do it forever.[2]

Belonging to a minority in the space you're creating can pose significant challenges to finding confidence in expressing your true self. Therefore, for example, a person with a disability, genderqueer, queer, neurodivergent or people of colour can often feel "different" or "othered" in audio and music production spaces. If you wish to delve into the experiences of some of these groups or learn more about finding confidence in your uniqueness, please refer to the recommended further reading at the end of this chapter.

As mentioned in Chapter 1, creativity can be explained by combining what we have learned and what our world perspective is. Would it not be boring if we all thought the same way? Our unique view of the world and our authentic emotions and reactions are what challenge the world and keep it moving forward. That is why art has always existed and will always be needed, whether that is in education, business, politics, or entertainment. Art has a great power to change things, move us, and make things go forwards, but for that to happen, we as creators need to say what we truly feel.

We will continue this conversation in the next chapter, discussing how the music industries truly affect our creativity and our creative confidence through factors such as genres, marketing, and social media. You will get further techniques for finding and maintaining the confidence in your own voice in the arts.

> Musician and Creativity Empowerment Coach Sara Belle comments:
>
> Belonging is part of the human condition. We want to fit in and feel "normal", seen, and heard. But once we throw the idea of "normal" into the mix, things get complicated. We now have a range of societal and institutional conditions against which to measure ourselves, our self-worth, and our ability to belong.

> The more we try to "fit in", the harder it can be to find our voice and feel that we can be our authentic selves. We develop a distrust for ourselves, dismissing the beauty and power we innately possess.
>
> But here's the trick, the more we tune into our own voice and let our intuition and self-trust guide us, the more we navigate towards the things that really resonate with us, the things that make our souls sing. Once we are in these spaces we find our sense of self, we find our tribe, and we begin to make a little community centred around the things we love, we find a sense of belonging.
>
> My creative practice has been instrumental in helping me find my voice and a place to belong. Once I decided to reconnect with making music, I discovered new ways to express myself and connect authentically with others. I started to feel like myself, trust myself, and hear my voice. I found a sense of genuine belonging.
>
> So, if you are feeling like you do not fit in, start listening to yourself and doing more of what you love. Don't be afraid to go alone, don't be afraid to try, don't be afraid to fail. Just keep listening to yourself and taking the next step in front of you. You will soon find your true tribe.[3]

## Practical Methods for Practicing Authenticity

**Authenticity List**
(Refer to Figure 3.3 for an example.)

- Write down when you feel the happiest.
- Write the moments you feel the most excited about life, in that moment and in the future.
- Consider where you experience these feelings, what you are doing and who is with you.
- This list can identify situations that might be toxic or unhealthy for you, making it easier to remove them from your life.
- With the aspects that make you feel good and optimistic about the future, try to see if you can increase these moments in your everyday life.

# I Am Too Different to Fit In 63

**MY AUTHENTICITY LIST**

○ I feel happiest when I am at home with my partner, my friends or by myself making music.

○ Whenever I think of a new music or art project, I get so excited that it makes me feel alive.

○ I struggle to show my music to Lisa because I really would like her approval, but she never manages to validate my effort. That never makes me feel great

FIGURE 3.3 Illustrated by Emma Holdway

## Notes

1 The School of Life. 2018. "The true and the false self." YouTube. Accessed 8 November 2024. https://youtu.be/A02Ucd6monY.
2 Rachel K. Collier. (August) 2022. Interviewed by Liina Turtonen.
3 Sara Belle. (October) 2021. Interviewed by Liina Turtonen.

## Additional Resources

- Burcaw, S. 2016. *Laughing at My Nightmare*. Square Fish.
- Buckle, B. 2023. "Lack of women and non-binary people working in music tech highlighted by new report". *Mixmag*.

- Cheryan, S., V. C. Plaut, P. G. Davies, and C. M. Steele. 2009. "Ambient belonging: How stereotypical cues impact gender participation in computer science". *Journal of Personality and Social Psychology*, 97(6): 1045–1060. https://doi.org/10.1037/a0016239.
- Disability Arts International. Three Adventures in Accessible Music Technology (AMT). www.disabilityartsinternational.org/resources/three-adventures-in-accessible-music-technology-amt/.
- Goodman, B. K. 2022. *You Gotta Be You: How to Embrace This Messy Life and Step into Who You Really Are*. Legacy Lit, an imprint of Hachette Book Group.
- Hendriksen, E. 2018. *How to Be Yourself: Quiet Your Inner Critic and Rise Above Social Anxiety*. St. Martin's Press, illustrated edition.
- Jordan, B. 2021. "Why aren't there more female music producers?" YouTube. https://youtu.be/2Ipb81z46kI.
- Joseph, S. 2016. *Authentic: How to Be Yourself and Why It Matters*. Piatkus, reprint edition.
- Mind Tools Content Team. Learning Styles. Mind Tools. www.mindtools.com/addwv9h/learning-styles.
- Nash, K. 2022. *Positively Purple: Build an Inclusive World Where People with Disabilities Can Flourish*. Kogan Page.
- Shetterly, M. L. 2016. *Hidden Figures: The Story of the African-American Women Who Helped Win the Space Race*. HarperCollins.
- Singer, A., G. Montgomery, and S. Schmoll. 2020. "How to foster the formation of STEM identity: Studying diversity in an authentic learning environment". *International Journal of STEM Education*, 7(57).
- Smith, S. L., K. Pieper, M. Choueiti, K. Hernandez, and K. Yao. 2001. *Inclusion in the Recording Studio? Gender and Race/Ethnicity of Artists, Songwriters & Producers across 900 Popular Songs from 2012–2020*. USC Annenberg. https://annenberg.usc.edu/research/hollywood-diversity-report/inclusion-recording-studio.
- Zwaan, J. "Josephine Zwaan". https://uva.academia.edu/JosephineZwaan.

# My Music Does Not Fit into Any Existing Genre     3.4

## Keyword

- musical genre

In the previous chapter, we discussed finding our voice as artists and ways to maintain our individuality in the arts. In this chapter, we will continue this conversation in a more industry-specific direction, asking how the modern music industry affects our creative choices as musicians. We will consider genres, expectations, management and representation, marketing, gatekeepers, and social media, and assess how these can affect how we make music, as well as affect our confidence.

## In Which Box Do You Belong?

When you tell a new person you meet that you are a musician, the first question they often ask you is: what type of music do you make? You might feel like answering with genres or referring to other artists. Forming this answer can be extremely difficult and can lead to an identity crisis.

What if you answer house music? Maybe they think you only make electronic dance music? If you say electropop, perhaps they think your music is very commercial? It can be impossible to define your repertoire in a couple of words, as it is connected to your identity. These questions make you quickly feel like every song you have ever made defines who you are and who you will be, and people will form their judgement of your music.

DOI: 10.4324/9781003194484-8

While the identity of a music genre is highly personal, why can it make us feel like there is a right and wrong answer to this question?

For instance, if you say your genre is house music, people may assume that you only make electronic dance music. On the other hand, if you say your genre is electropop, people may think that your music is very commercial. Defining your music genre can feel like defining your identity, and it is not always easy to put into words. It is natural to feel like people will form their judgement of you and your artistry based on the genre. However, whatever art we might do is complex and personal. So, why do we often feel like there is a right or wrong answer to this question?

Just like we use fashion to express our identity, we also use music to define our personality. Hargreaves, Miell and McDonald describe this well:

> Music can be used increasingly as a means by which we formulate and express our individual identities. We use it not only to regulate our own everyday moods and behaviours, but also to present ourselves to others in the way we prefer.[1]

They continue with the statement: "Our musical tastes and preferences can form an important statement of our values and attitudes, and composers and performers use their music to express their own distinctive views of the world."[2] The idea that artists choose their genre based on their creative expression seems obvious, but in reality, this is not always the case in our modern world. Despite the assumption, many artists may not have the freedom to fully express themselves due to external factors such as marketability and commercial success.

Modern musicians, especially music producers, use genres as their main guideline, not their emotions. The focus is to make music that fits a specific Spotify playlist, matches the established social media brand, or ensures each single released is in the same genre so that the social media platform algorithm does not get confused. The genre has become more than an identity or result of an expression. It is a business model. Therefore, as we need to define our musical genre very early on, we must also make sure we choose the right one. And this is where the confusion happens; genres have a substantial identity value, so how do I know who I am and what I believe? If I make pop music, does it seem like I am selling myself or do I need to make rock music so others take me more seriously? Am I allowed to change my direction?

As mentioned in the previous chapter, we all have the need to belong one way or another, and this comes from the fear of being left out or not liked. Many communities are explicitly directed towards specific points of interest in the music and audio industries. If you like modular synthesisers, you might join online groups featuring conversation and tips on the topic, and this is the same for every other program, tool, and genre. But as much as these communities can be encouraging and inspiring, there can be an underlying toxic influence, which often comes in the form of gatekeepers.

As mentioned in Chapter 3.2, the music and audio spaces showcase plenty of opinions from peers, students, and educators who might have lots of confidence in their skills and music. But in these spaces, the confidence does not always present itself in an encouraging way. Instead, it can come across as *conscious* or *unconscious boasting*. This means that some individuals may express their opinions strongly on what they believe is right or wrong in the topic discussed. They may also resort to harmful or bullying behaviour towards other creators representing opposite genres or musical styles. This type of gatekeeping can be particularly daunting for someone still learning and trying to find their footing in the creative process. It may even make them doubt their abilities and question whether their music is good enough.

As much as genres and tools can have standards that are good for everyone to learn, in the end, they are just guidelines. These communities should always only inspire and teach us encouragingly to experiment and find our way in the industry without damaging our creative confidence. We will continue discussing the *right* and *wrong* in the audio and music industries in Chapter 3.7.

## Consistency Is a Tool for Finding Your Unique Voice in Art

There is an excellent metaphor for finding a sustainable and healthy career in arts while finding and maintaining your originality. It is called the Helsinki Bus Station Theory.[3] The theory by Finnish-American photographer Arno Rafael Minkkinen describes our artistic development in the industries in a practical yet realistic way. He illustrates the beginning of a career like a bus station, a space with many platforms and buses with different numbers. Each of these numbers then heads out of the city centre, first stopping at the same bus stops, until coming out of the centre and heading in different directions.

This theory is a metaphor for our career in the arts. The bus station is where everyone is at the beginning, the bus numbers are different skills and directions you can take in art, and the bus stops are the years you spend practising your craft (refer to Figure 3.4).

When we get on our bus and start to study the craft we choose, we notice that others from the different buses will be stopping at the same stops, or

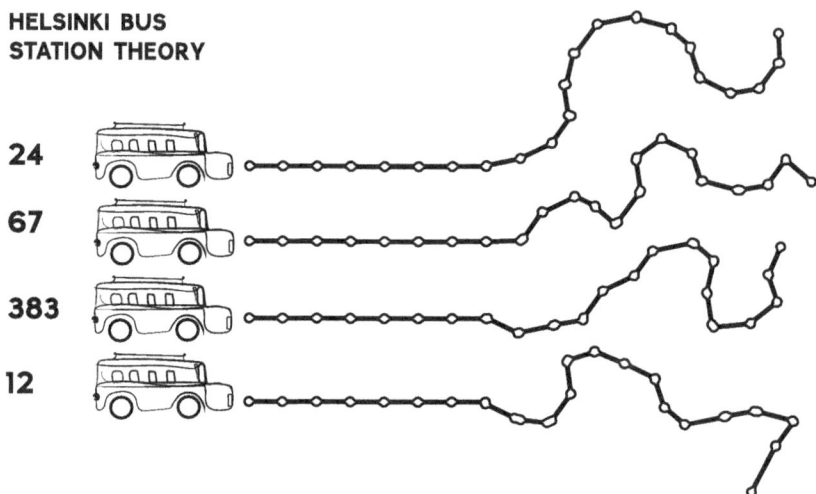

FIGURE 3.4 Illustrated by Emma Holdway

you might even feel that your bus is very crowded. You are all learning similar skills, and the comparison to others is heightened. This is where Minkkinen says most of us take a taxi back to the central station and look for another bus to hop on. We can find happiness in learning these skills, but as the bus station and buses are crowded, it can be tough to find your voice. But as said, the bus stops reflect years, so what if you stuck with one bus line and saw what happened when it leaves the city?

If you stay on the bus, this is where separation from others starts to happen. Both in yourself and how your work is perceived in the industry. Minkkinen explains:

> It's the separation that makes all the difference. And once you start to see that difference in your work from the work you so admire – that's why you chose that platform after all – it's time to look for your breakthrough. Suddenly your work starts to get noticed. Now you are working more on your own, making more of the difference between your work and what influenced it. Your vision takes off. And as the years mount up and your work begins to pile up, it won't be long before the critics become very intrigued, not just by what separates your work from a Sally Mann or a Ralph Gibson, but by what you did when you first got started!
>
> You regain the whole bus route in fact. The vintage prints made twenty years ago are suddenly re-evaluated and, for what it is worth, start selling at a premium. At the end of the line – where the bus comes

to rest and the driver can get out for a smoke or, better yet, a cup of coffee – that's when the work is done. It could be the end of your career as an artist or the end of your life for that matter, but your total output is now all there before you, the early (so-called) imitations, the breakthroughs, the peaks and valleys, the closing masterpieces, all with the stamp of your unique vision.[4]

As we discussed previously in this book (Chapter 1), creativity happens at the crossroads of what we have learned and how we see the world from our perspective. And what Minkkinen says supports this theory. In the beginning, we might be on the bus with others, crammed together for years, competing for space and visibility. But while doing that, we learn not only skills and the craft, but with time we experience life and figure out our relationship with it. When the buses leave the centre and travel to their paths, we start to combine our learned skills and viewpoints in a new unique way, not from the perspective of others, but our personal view of ourselves and the world around us.

So, as frustrating as it might be, remember that life, art, career, healing, or whatever you are doing, is not a sprint but a marathon. If we feel lost in our artistic voice, or feel like things are coming to us harder than others, it does not mean we are more lost than others. It just means we might still be sitting on a bus closer to the bus station. We only need to keep our bums glued to the bus seat, be patient in our progress and have faith in our journey out of the centre.

## Embrace Your Individuality

### Author's Experience

Before I learned music production, I composed most of my music using guitar, piano, and vocals. The options for genres were limited to how these sounded together. But learning DAWs, synthesisers, and other electronic tools, I was suddenly open to thousands of new genres, some I had never even heard of. This confused me about my identity as an artist and my place in the music industry. Personally, I never felt comfortable fully delving into the indents of a *singer-songwriter girl* or *electronic house producer*. I have always wanted to be me, but that is not, unfortunately, a genre option when submitting a song to radio. This was my struggle for years until I decided to do something to change the way I felt artistically.

I started creating YouTube videos in 2019 and posting every week for several years not only trained me to become a more skilled producer but also taught me more about my personality than I ever thought. For example, through the comment section on YouTube and putting my moderately unfiltered self on the internet, I learned that some people find me annoying and too loud. In contrast, others see me as inspiring and even funny; some things about my personality I was not fully aware of before. And aside from the comments, being consistent with creating content weekly for almost four years taught me about my values, what I see as success, why I make art and who I want to be as a person and an artist.

One of the critical moments in my music career was when I made the song *No, I'll Do It*. It came after years of focusing only on production skills, not releasing music for years and having a severe artistic creative block. I wanted to create it without any genre expectations, and I also filmed a music video where I explored the humour and freedom of DIY-style video production. *No, I'll Do It* came from anger and tiredness in the censoring elements of the world affecting our creativity. This project focused only on what I felt and wanted to express as an artist, without a single thought for the music business, likes, sales, genre, or the opinion of my friends, family, or followers. I was extremely nervous about releasing it, fearing people's reactions, because for the first time, I put my vulnerable self out on the internet. But interestingly, the song did better than any other song I have ever released.

When thinking about the Helsinki Bus Station theory, I feel like I am still on the inner-city bus, learning and navigating where I wish to be with my art. But after consistently working for years, I can now see my journey forming behind me. Slowly, one year at a time.

There will always be people who might not like what you do, but there will be plenty more who will support you, especially when you stay true to your perspective and vision.

In Chapter 4, we will talk about plans and workflows, which will help you practically visualise your journey, help you find direction when you feel stuck, and find out what your goals and measurements of success are. All of these aspects guide us towards figuring out our sound as artists and the healthy ways to maintain it in our careers.

## Three Exercises You Can Practise When You Feel Lost In Your Artistic Identity

1. **Practise finishing music without judging it before it is finished.** It is common to have high expectations about the final outcome of our music before we even begin creating it. For example, you might imagine your track becoming the next big hit in a particular genre and start picturing scenarios and stories for the outcome long before the piece is ready. However, this can lead to unrealistic expectations as creative processes are often unpredictable, especially if you are still in the early stages of your production journey. Here are some tips to help you with this process:
   - Learn to finish music and tracks, even if you dislike them. This is how you learn the habit of completing creative projects. And the more you finish music, the more you will make pieces you truly love.
   - When you do not like the music you are making, use it as a chance to solve a puzzle and learn about yourself. What are the aspects you do not like about it? What sounds inspire you on other people's tracks which are missing from this one?
   - Remember that something you do not like now does not mean it is bad or that you will never like it. So, who are you to say if the tracks you are making now are good or bad? Let them happen, and do not be so hard on yourself and your creativity.
   - Finally, this is the most important tip. Decide what you wish to do with your music only after you have made it. This is the time to decide which ones you want to release and share with the world.
2. **Use the power of an artist name to separate your multifaceted creativity.** Often, we feel baffled when we are asked to define ourselves with one box and describe all our music just with one genre. This confusion is natural and happens because we are humans, and, inherently, we are way more complex than that. If you are feeling this, try out some of these techniques:
   - Use artist names to separate your own identity from the commercial one. For example, my given name is Liina Turtonen, which I like using for work that does not depend on boxes or genres. Alternatively, I use the artist name LNA for releasing chill house music. On top of this, I have a secret techno artist name, which I want to keep private to release dance music I do not want to feel pressured about. It is just for me and only for fun.
   - Use your different names to separate the expectations you have for the outcome. So, for example, if you wish to make music your career, treat the artist name dedicated to that as a business. This name is then

a more structured and curated piece you show to your audience. For me, LNA is this. I still show the authentic side of me, but only how it fits the LNA brand and goal. Doing this manages your expectations, provides structure for development and also gives the brand a clear vision of who you are and what you can offer. Consistency and definition are what all social media accounts and streaming services want for growth. But as I have other artist names, I am not putting all my identity into one box. Instead, I gain more by compartmentalising both my business goals and artistic identity.

## Notes

1 David Hargreaves, Dorothy Miell, and Raymond Macdonald. 2002. "What are musical identities, and why are they important". In Raymond MacDonald, David Hargreaves and Dorothy Miell (eds), *Musical Identities*. Oxford University Press.
2 David Hargreaves, Dorothy Miell, and Raymond Macdonald. 2002. "What are musical identities, and why are they important". In Raymond MacDonald, David Hargreaves and Dorothy Miell (eds), *Musical Identities*. Oxford University Press.
3 Arto Rafael Minkkinen. 2004. "Finding your own vision." *JamesClear.com*. Accessed 8 November 2024. https://jamesclear.com/great-speeches/finding-your-own-vision-by-arno-rafael-minkkinen.
4 Arto Rafael Minkkinen. 2004. "Finding your own vision." *JamesClear.com*. Accessed 8 November 2024. https://jamesclear.com/great-speeches/finding-your-own-vision-by-arno-rafael-minkkinen.

## Additional Resources

- Csikszentmihalyi, M. 1996. *Flow: The Psychology of Optimal Experience*. Harper Perennial.
- Gilbert, E. 2015. *Big Magic: How to Live a Creative Life, and Let Go of Your Fear*. Riverhead Books.
- Pressfield, S. 2002. *The War of Art: Break Through the Blocks and Win Your Inner Creative Battles*. Black Irish Entertainment LLC.
- Struthless. 2020. "Advice for people who feel like their art isn't unique". YouTube. https://youtu.be/vtt1ycq2rVI.
- Turtonen, L. 2023. "I Was Just Thinking, unlearning harmful self-beliefs". YouTube. https://youtu.be/1CW4qGpAEUA?si=JRnBxvsaUiV_vMBN.

# I Do Not Know Music Theory so I Cannot Learn Music Production

## 3.5

### Keywords

- music theory
- audio software

### Music Theory vs No Music Theory

Have you ever tried to explain to someone what music production is? This multifaceted term includes being an artist who learns new techniques, captures performances, and freezes them in time. They are the conductor of a personal orchestra, the director, and all of the players simultaneously. It is pressing buttons, turning knobs, and hearing sounds in immense detail.

Anyone listening to your description thinks you are superhuman; someone who has talent above standard human capability, a musical hero with knowledge in music and composition, songwriting, technology, audio, playing instruments, performance, and singing. All of these topics are massive, and focusing on learning just one of them can take years. So it is not a surprise that the expectations associated with this title can cause insecurities for so many of us. And, among all the technical knowledge, one of the skills expected is music theory.

Music theory helps a person to contextualise, analyse, give sound structure, and widen a musician's scope to improvise in many different styles. Studying music theory can be compared to learning a new language; it is the grammar of music.[1] It can help the musician understand what they are creating, be consistent in their creativity, and expand their vision outside of their mind using analytic thinking. Like any science or theory-based subjects, knowledge infinitely expands what we already thought could be possible.

But when talking about theories related to creativity, we should ask ourselves: does knowledge always have an expanding quality, or can it also have a reducing effect when applied at the wrong time or by pressure? As mentioned in the last two chapters, the audio and music industry spaces affect our confidence and, therefore, our creativity. This means that the aspects of competition, group hierarchy, or structural elitism can affect how approachable and accessible we consider music production to be.

Originally, music theory had its roots in classical music and composition, and was associated with higher education. Therefore, it has been, and in some instances still is, a skill often associated with a person's socioeconomic background. But with technological change, music-making has become available for everyone, regardless of their circumstances. This means that you can become a music producer without a degree, knowledge of music theory, or references for musical history and culture. For the people who value traditional music education, this change can be hard to accept, as they consider music theory highly valuable for creating complex and exemplary art. But for some, this change has brought freedom to create on their own terms and without the bonds of traditional constructs.

During the past few years, there has been a noticeable increase in audio plugins, physical interfaces, and other tools that let people produce songs without knowing music theory. Most audio software now allows you to limit MIDI notes into the selected scale to create chords and melodies without knowing what notes belong to them. Similarly, you can audition and convert samples directly into the scale with a couple of clicks. The plugin and device developers constantly bring new tools to create chords, melodies, arrangements, and even lyrics for faster workflows.

The discussion of whether using loops, samples, AI programs, and assistive tools are *cheating*, is very current. This forces us to consider the amount of creative input we need and whether these copy and paste techniques constitute as art.[2] In music production lessons, it is common to see students wondering if using the DAWs default sample library, presets, and generative tools are appropriate. Legally, you have the right to use these royalty-free samples in your music, but the perspectives of this conversation can depend entirely on a person's background and musical values. But no one can define if a song made partly from samples is *better* or more *authentic* than a song recorded from scratch with an acoustic guitar, because there is no right or wrong way to do art.

In the end, how do AI apps and traditional music theory knowledge differ? They both assist the creator in expanding their vision, help them to be more consistent in their practice, and make their workflow faster. Again, there is no right or wrong in this, but there is one simple favour that we could all do for our collective creativity: stop judging each other's process by what they choose to use to assist them in their creative process.

### Author's Experience

I moved to the United Kingdom from Finland when I was 21 and began a university degree in Commercial Music. After settling into my course, I noticed that I was one of the only people in the class who knew music theory. For someone coming from the Finnish educational system, this was a shock. Music theory is held in remarkably high regard in my home country, and it would be tough to get on a university-level course without at least knowing the fundamental principles of it. As I had always struggled to learn music theory and, as such, would have likely never managed to get into a university course in my home country, this new experience seemed exciting to me. Maybe finally, I could feel like I belong.

During this degree, I discovered a new point of view. The students who had played music all their lives and learned without reading notes and understanding time signatures had a unique freedom in their creativity. They had confidence in their expression in music without getting stuck thinking about chord progressions or scales. Instead, they trusted their ears. I felt like they were the people who had the most presence in practice and enjoyed the playing. This playfulness of their music-making was something I began to be very envious of, but at the same time, it inspired me.

This university experience made me see the freedom in a different type of musicianship and I wondered if my conservatoire background had limited me from breaking the rules of music because I was scared to make a mistake or play by ear. Later in life, I found a balance between these two worlds. I am grateful for my mother for putting me in the music theory classes as a child, as now, in my work as a music producer, I have noticed that even knowing the basic level of theory has helped me hugely. But at the same time, I struggle with creative blocks, wondering if I could create better music with more advanced music theory knowledge. In fact, my brother is also a musician, and out of the two of us, he has always been the one who knows more about music theory.

> Sometimes, I feel like if he had produced the songs I made, they would be more complex, exciting, and possibly more successful. But with this comparison, I am just becoming more insecure, which contributes nothing to my music-making, only limiting my creativity with self-sabotaging thoughts. Therefore, I can sign up for a music theory refresher course or stop stressing about it all together, and buy a plugin to assist me.

These topics and themes will be discussed in more detail later in this book. In the next two chapters, we continue the conversation about music equipment and the pressures connected to gear fetishism, as well as the concept of right and wrong in music-making, and what makes us feel like these rules of music exist.

## An Exercise for When You Feel Overwhelmed About Not Knowing a Specific Topic in Music Production

### Create a Map That Helps You Navigate Topics In Music and Audio

As mentioned, music production is a multifaceted skill, and sometimes it can be overwhelming to learn it, especially when you do not know all the things that might benefit your practice. That is why it is good to dissect all the subjects into categories, which helps you create an action plan according to what you are interested in and what your goals are.

Figure 3.5 lists some areas of music production that represent key categories for knowledge and technique. (This list is merely an example; please proceed with topics relevant to you.) You can note down the subjects you want to explore further, and then specify the ones for which you prefer to find assistive tools. Doing this helps you to feel less overwhelmed, reminding yourself that it's okay not to know everything and encouraging a focus on the techniques and tools you currently possess, rather than those you have yet to acquire.

Remember, you can learn and search for assistive devices simultaneously. Perhaps starting with a plugin or tool to help with a subject at the beginning can alleviate the pressure of learning. Additionally, even after mastering a topic, you might still want to use tools to enhance the efficiency of your workflow.

# I Do Not Know Music Theory so I Cannot Learn Music Production

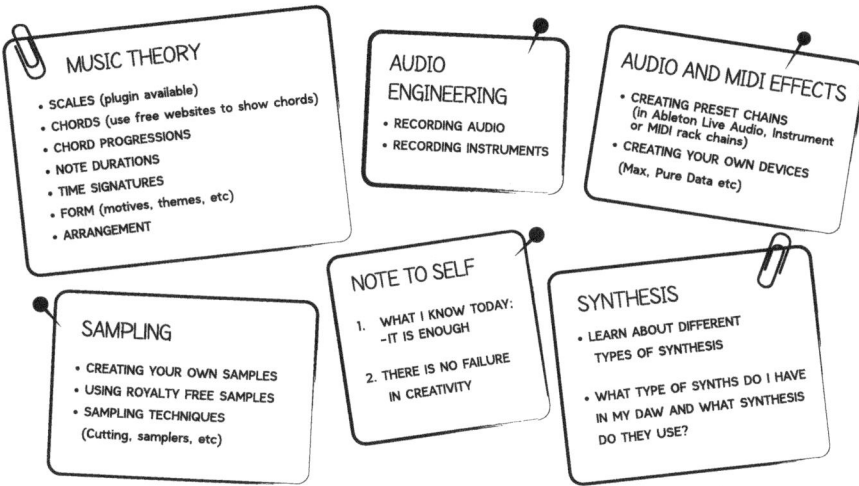

FIGURE 3.5 Illustrated by Emma Holdway

## Notes

1 Eytan Jun. (25 July) 2017. "Importance of music theory". *Liberty Park Music*. Accessed 15 January 2024. www.libertyparkmusic.com/the-importance-of-music-theory/.
2 Paul White 2008. "Are you cheating?" *Sound On Sound*. Accessed 15 January 2024. www.soundonsound.com/techniques/are-you-cheating.

## Additional Resources

- Ableton. Learning Music (Beta). https://learningmusic.ableton.com/.
- Lightnote. www.lightnote.co/music-theory/sound-waves.
- Plugin Boutique. The Best MIDI & Music Theory Plugins. www.pluginboutique.com/articles/1782.
- Mineo, L. (15 August) 2023. "Writer, animator, architect, musician, and mixed-media artist detail potential value, limit of works produced by AI". *The Harvard Gazette*, n.p.

# I Do Not Have Enough Money and the Right Equipment

# 3.6

## Keywords

- Ableton Live
- gear

In the last chapter we mentioned how audio spaces, the amount, and the price of the equipment can activate several insecurities in us: feeling like we are not enough or that our economic situation will set us back from having the chance to create successful careers in the music industry. This chapter will look into the insecurities that audio, music, and production equipment can create in us and how we can navigate our artistry and goals, according to our economic situation.

## The Equipment Is There to Help Us in Our Vision, Not Carry It

How do you feel when you see a professional music producer or an engineer in a big recording studio, while the place you create in is a home studio with budget equipment? Comparing the number of tools and how much money has been spent on their equipment can create an illusion that the person might have a better chance of making more popular songs than you. This comparison can make you feel inferior, have feelings of insecurity and doubt your chances of creating professional standard music. But this scenario is not as black and white as it seems.

Creativity always comes from our mind and experiences, and the equipment is there to help us in our vision. In the same way a painter uses brushes to put the paint onto a canvas, we might use microphones to capture audio or plugins to modulate our signals to be aesthetically pleasing.

In the audio industry, it is easy to believe that you need to have specific devices to make better or more professional music. Social media is full of advertisements for assistive tools and plugins that promise faster and better results. Maybe the DAW you are using already has all the tools you will need? Perhaps you do not need to spend more money to purchase the latest release of the shiny new plugin to be more professional?

While I was still doing my master's degree in music production, these were the questions I was asking myself a lot. I was on a very tight budget and did not have money to buy almost any tools or plugins. The only things I was able to invest in at the time were a new laptop, a pair of headphones and Ableton Live Standard. At the time, I felt it was impossible to make *radio-ready* quality music, as I did not have as much equipment as my fellow students. But interestingly, my lecturer soon taught me this was not the case.

We did not talk about music software, equipment, or plugins on the course. We were not told to use specific tools or buy a particular brand. But what we focused on the whole year was understanding how audio works. This meant having complete knowledge of an audio signal from a physical perspective, how effects are made from a mathematical viewpoint, and what happens inside a microphone when it captures sound in a specific room. We used free recording software Reaper to capture sound and code our plugins using JavaScript coding language.

Initially, this approach was weird for me, as I had become used to the main focus of education being on learning the best gear to make music. Still, it was about learning the science of audio to enable us to artistically do whatever we wish, with whatever tool we choose. I understood that once you know the basics of audio physics, any plugins, synthesiser, DAW, or device make sense straight away.

For example, I did not feel like I needed to feel insecure that I could not afford the latest plugins, as the default devices in Ableton Live were enough to do a professional level of music. For production and mixing, you need tools to work with dynamics (selection of compressors), the frequency spectrum (equalisers), the colour and harmonics (saturator and distortion tools), and stereo spectrum (panning and other stereo utilities). For the first time, my goals with production and mixing were led by conscious decisions of my artistic vision and not by the tools.

Only with time and a better economic situation have I expanded my library of plugins and tools. But I only get them when I genuinely feel they give something extra to my ability to work with audio, or enhance my workflows. Audio equipment is a fun and exciting thing to have, as long as you remember they do not make your song professional, excellent, and compelling. You do.

So the next time you feel insecure about the tools you have to work with, remember that the equipment does not make a musician, and the amount of expensive gear one might have in their studio does not make them more knowledgeable or better at music than you. I have additionally collected a selection of extra resources to look into at the end of this chapter. These articles are an inspiration for anyone working in music production and feeling any type of gear-related anxiety.

## Gear Fetishism

It is typical to love gadgets, things that look and feel exciting, but with this, we can hide some of our insecurities regarding our creativity. The number of instruments and tools can feel intimidating in a creative situation. Maybe you do not yet know how to use all of them or perhaps there are so many creative ways to approach your workflow that it is hard to know where to start. *Gear fetishism* is the obsessional need to collect and own the newest toys and tools and build the best-looking studio.[1]

A comment under one of my videos, on my YouTube channel LNA Does Audio Stuff, summarised the problematic side of gear fetishism:

> My problem is finding the motivation to keep going. I calculated how much I've spent putting together and building my home studio, and I honestly could've bought a Tesla. Now I just sit in my producer's cockpit and look at my mix console, synthesisers, outboard gear, monitors, acoustic panelling, LEDs and giant 4K display hanging out in front. Then like a cat, I decided to hop down and lick clean myself on my bean bag chair. I think most would give their right arm for my setup, but I'm now considering building a camper van. Motivate me!!![2]

Sometimes, social media content creators, like myself, are part of the problem. It is not often apparent to viewers that many of the tools and instruments in the videos are sponsored or gifted by the brands, creating unrealistic expectations about how professional music studios should look like. Sometimes it can be hard to figure out the differences between what the creator can realistically afford and which items they truly love and use in their creative workflows. It is natural to think that someone with many synthesisers and instruments must be very good or professional. But perspective can deceive us. We can start to value their achievements through the image they present while unconsciously putting the worth of our creative efforts lower.

The key is to find confidence in your creativity, believing that your ideas are enough for whatever your goals might be. No expensive and well-equipped studio can replace a freely creative mind. Therefore, whatever equipment you can get your hands on, be mindful of what you plan to do with it. For example, let us say you currently have a smartphone but lack any other music production equipment. Nowadays, the quality of the free software you can create music with is impressive. One free app is called BandLab. It is browser-based music-making software that gives you all the tools to create good quality productions and mixes. Yes, it has limits on what it can do, but it has enough features to freeze your creativity into time.

In this case, what is there to prevent you from creating a hit song using just your phone? Most of the time, the limitations to what we can make are more in our heads and imaginations than in the tools. By recording, and using samples, presets and effects on even simple devices, you can be surprised at the fantastic music you can create, just as long as you have fun and trust in the process.

Sometimes limits are enhancing factors to your creativity. Focusing on buying new gadgets and tools can possibly distract you from your creative process. Learning to use new equipment can take a while, making it easy to have self-sabotaging thoughts when you cannot use them in your music. You might start feeling overwhelmed and questioning yourself and your skills while wanting to quit the session. In these moments, creativity becomes more about the devices and less about what is essential: having fun, creating art, and expressing yourself.

In Chapter 4, we will discuss further how technology and the audio community affects our creative workflows and confidence. You will be introduced to practical step-by-step guides to help you overcome any negative feelings and insecurities you might feel due to the equipment.

> **EMBRACE OR CREATE LIMITATIONS**
>
> An exercise to help you shift the focus from your audio gear and tools into your creative process.
>
> 1. Limit your time (try time blocking, which is explained in Chapter 4). This can be a 30-minute, 2-hour, or even a 1-week time limitation, in which time you challenge yourself to create a piece of music.
> 2. Choose only a limited amount of equipment to help you with your workflow. This can be a phone app or even one instrument or feature in the music-making software.
> 3. Remember not to think about the outcome while creating, fully focusing on the process.
> 4. Whatever your will create, let it happen. There is no failure in creativity, so if the outcome is something you would not publish, that is okay. The more we create, the more we make art that we like, and the other material that comes is part of the process.

## The Audio Industry Is Expensive

The audio and music industry is indeed an expensive hobby and career to have. To be able to start production, you do need to have access to a relatively good computer and music-making software. Even a recording bundle with an interface, microphone, and headphones can be too dear for many, making it hard to even consider pursuing it further. If you wish to go and study music production at a university, depending on the country, the courses can be thousands of pounds, while still needing some of the equipment at home. This is why music production and audio industries are not accessible for many to enter due to financial constraints.

The psychological and cognitive science research journal published by PNAS in 2019 discusses the reasons for why people from lower-socioeconomic backgrounds end up pursuing STEM (science, technology, engineering and mathematics) careers. They mention how the anxieties and insecurities around STEM subjects start even as a child in school: "The numerous structural barriers between lower-income students and STEM preparation in high school – such as neighbourhood factors, types of schools available, systemic prejudice based on social class – create the sense that the only ways to help involve needed large-scale changes to schools and society."[3] The barriers can quickly leave people stranded with their insecurities and financial limitations without considering a STEM career such as audio could be a possible career path.

While the audio industry may be filled with biases and hardships caused by socioeconomic status, these do not need to be indefinite barriers to moving forwards in your aspirations. Some factors of the industry have started to recognise these barriers and create opportunities for support to enhance the accessibility of entering the audio industry.

In increasing numbers, schools create supportive structures for children from different socioeconomic statuses, giving them access to equipment and free after-school activities. Many organisations also offer free access to courses for young people who wish to learn technology or audio. Similarly, in university education, you can find funding to cover the education fee and options to borrow computers and other equipment while studying. For example, in the United Kingdom, you can find more information from organisations such as UK Youth, Shine, Action for Children, Brighter Sound, and many others.

## Notes

1. Alex Annets. 2015. "Masculinity and gear fetishism in audio technology community discourse". Anglia Ruskin Univeristy Researc Online (ARRO). Thesis. Accessed 15 January 2024. https://hdl.handle.net/10779/aru.23758530.v1.
2. LNA Does Audio Stuff. 2022. "How to make techno track in 10 steps – Full song from start to finish". YouTube. Accessed 15 January 2024. https://youtu.be/8om2uak_FKU.
3. Christopher S. Rozek, Gerardo Ramirez, Rachel D. Fine, and Sian L. Beilock. (29 January) 2019. "Reducing socioeconomic disparities in the stem pipeline through student emotion regulation". *Proceedings of the National Academy of Sciences*, 116(5): 1553–1558. https://doi.org/10.1073/pnas.1808589116.

## Additional Resources

- Bates, E. and S. Bennett. 2022. "Look at all those big knobs! Online audio technology discourse and sexy gear fetishes". *Sage Journals*, 28(5). https://journals.sagepub.com/doi/abs/10.1177/13548565221104445.
- Dacombe, M. 2023. Artists Who Made Global Hits with GarageBand. *Medium*. https://medium.com/@certifiedmish/artists-who-made-global-hits-with-garageband-32314c0861e1.
- Hoby, H. 2012. "One to watch: Grimes". *The Guardian*. https://amp.theguardian.com/music/2012/jan/29/grimes-boucher-one-to-watch.

- Jordan, B. 2023. "I'm done w/ gear videos & reviews". YouTube. https://youtu.be/FEO0NVPyED4?si=sH2IFE3Xax5zTWQu.
- Sorcinelli, G. 2016. "From GarageBand loop to Grammy Award: A look back at Rihanna's 'Umbrella'". *Medium*. https://medium.com/micro-chop/rihannas-grammy-award-winning-umbrella-is-a-garageband-loop-3e1430446363.
- WIRED. 2017. "How the internet's Steve Lacy makes hits with his phone". YouTube. https://youtu.be/E6BxAtc5cd0?si=dQCpvh2Cs_48H_Gz.

# There Is a Right and Wrong Way to Make Music

3.7

## Keywords

- equaliser
- compressor
- sidechaining
- The Quartet of Creativity

It is very common to feel insecure about whether the music production techniques or plugins that we use are correct for a particular genre. The social media and YouTube communities can make us question our knowledge and whether the way we do things is *right*. Therefore, this chapter will discuss how we know what is *right* and *wrong* in our music production and music-making, and how we can overcome the insecurities this can cause us during our creative processes.

## Science Is Right, and the Rest Is Just Opinions

We all know the feeling when you go to a look for a YouTube tutorial, and the teacher gives direct statements such as: compressors always go after equalisers, or you should always sidechain your kick and bass. Afterwards, you see another teacher say the opposite, and now you feel confused about what you should do to get the best result possible. But what if I would say equalisers can go before compressors and you do not always need to sidechain your kick and bass because it depends for what purpose you are using these techniques. Watching someone you look up to, who presents themselves as a person with great knowledge of the subject, can affect what you believe is *right* or *wrong*.

DOI: 10.4324/9781003194484-11

## 86  Creative Confidence and Music Production

People often look for shortcuts to better results and guidelines when they struggle with understanding audio concepts. But the issue with teaching and learning audio is that most of the time, there is no right answer to anything, as it *depends*. Just as there is no right way to put paint on canvas with a brush, there is no correct way to create sound. Perfect guidelines to say how much compression someone should apply does not exist, as it entirely depends on the signal and the purpose of your use. Music production is a confusing art form as so much of it uses science as a tool, but in the end, the results are about the artist's taste. A producer, mixing engineer, or mastering engineer's job is not to make something *correct*, but to make it perfect for the song. Yes, there are industry-standard guidelines on what is expected, but inside these rules, the final aim is always to do art.

As drag queen RuPaul famously says, "You are born naked, and the rest is drag".[1] The same applies here. The audio science is correct and based on laws of physics, which is hard to argue against. But on top of that, everything else you do is your self-expression and artistic vision (refer to Figure 3.6).

In conversations with peers, it is often mentioned that with more knowledge and better active listening skills, the less you focus on equipment, the more you trust and value your ears as the main tools. As a beginner, it is easy to think you will get the best results by doing all the *correct* techniques. But as it is good to widen your method repertoire, sometimes the simplest approach can give you the best results. But you will only learn to know this with practice and by trusting your ears.

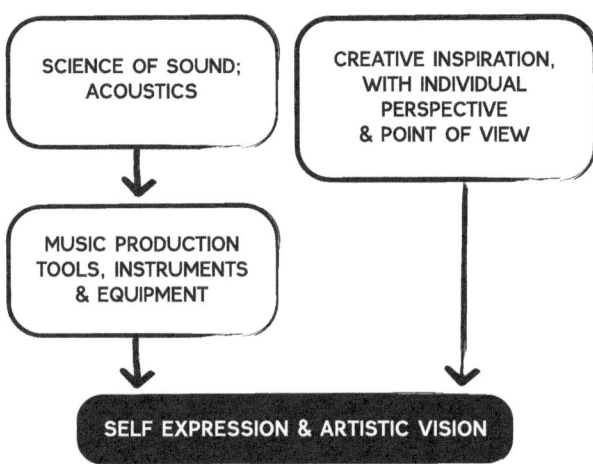

FIGURE 3.6 Illustrated by Emma Holdway

The next time you question whether your technique or approach is correct, try practising these steps:

1. Learn how audio works. Some key concepts to understand are signal shapes, frequency and stereo spectrum, and the dynamic range.
2. If you wish to learn a specific technique, try to identify the concepts relate to it. This means, if you, for example, wish to learn how a specific synthesiser works, dive deeper in your research into each function. For example, you could learn what synthesis the synthesiser uses, then the basics of different synthesis or how the oscillators, envelopes, and filters function.
3. Whenever someone states an instruction as a fact, consider whether this is just an artistic opinion or factual information about the physics of audio or how a specific tool works. For example, saying that LFO is a low-frequency oscillator that creates a pulsing effect when applied to a signal is a scientific fact. While parallel processing every time with dynamic signals, such as vocals, is just an artistic opinion.
4. Remember that tools and techniques are not the focus point of being an artist. They are just there to assist our creativity. So it is good to take time to separate the science from the your expression. If you feel overwhelmed by the learning and the tools, take a step back and focus on having fun without rationalising your actions.
5. Sometimes receiving our goals needs *correct* techniques related to the style or genre. For example, if you wish to produce a deep house track there are characteristics to this style that identify it from other genres. In this case approach the project with caution and try compartmentalise your artistic voice and the goal. Research first all the necessary techniques you need to know for this goal, proceed to apply them systematically. After, consider if you are happy with the result, and if not, use it as a template to express your creativity freely without any rules. This technique is great when you wish to create music in a specific genre, while making it unique to your taste.

## Are You a Real Musician?

You hear many conversations about *right* and *wrong* in music, and being a *real* musician, especially online, in music forums and channels. The question of what we consider as musicianship depends on what environment we were raised in and how that moulded our values in life. For some, being a musician means they can earn money with it, while others consider formal education defines a musician. In some cultures, everyone is a musician, and music is a way to relax and interact between family or a community.

Not long ago, participating in music actively was part of most people's lives. For example, the United Kingdom has a long history of community music where people would regularly play and sing together in churches, homes, or pubs.[2] However, this has seemed to disappear from modern Western society. Nowadays, music is often either a talent you were born with, a skill to showcase in a talent show, or mastery of an instrument that has been learned often in higher education. What can make the modern views of music-making and the music industry problematic is its effect on our creative confidence. If we value music through talent, merit, and skill, it sends a message that only a few people can do it by the standards the world expects from us.

There are clear benefits of community music practices for children and older people. Music is part of us growing up, learning cognitive, memory, and motor control skills.[3] And for the elderly, music can reduce memory loss and create activities that have mental health benefits. So, where does this leave everyone else? We consume music, possibly every day, through media, entertainment, and streaming, but for some, taking part in making it is almost an impossible idea. The only time people sing is possibly in the shower or drunk in a karaoke bar.

And why is this important to consider when we discuss *real* musicianship? It is the unrealistic bar that has been set for music-making. It seems as if there is no space for music that is made just for enjoyment, fun, and connection with others. As artists, we need to improve and be better continually, so we constantly reach for something bigger and better instead of finding comfort in our mediocrity. It feels like our art, and we, as artists, cannot be flawed, raw, or imperfect.

No wonder we might feel disappointed in our efforts to make a pop song when the skills and equipment needed to make a radio-quality hit take years to collect and master. It is also no surprise that we might feel like there is a *right and wrong* way to create music if we are not allowed to experiment and make songs without the impact of others. If we attempt to make something that we feel is truly amazing, but it does not reflect a style or standard of the music industry, it must be wrong. This is, in essence, contradictory to what art is often about and what makes it human.

In Chapter 1, we discussed The Quartet of Creativity, where it is explained how the space around us affects our creative decisions. The same theory applies to this topic. If we genuinely believe there is a failure in art, we start to limit our authentic selves and what we wish to express, and how we want to tell our stories. Finding the way to your true self is the key to a consistent and healthy career as an artist. But as we have discussed in the past three chapters about authenticity and the music industry, not following the trends and standards of others can make your career prospects limited. That is why knowing there is no right or wrong way to make music can be your compass inside these trends, guiding you to the goals you wish to achieve, but on your terms.

Consider what are the standards you value in music, and where do you find beauty in it? The way a child can belt out an alphabet song without fear of failure, enjoying the experience without caring if they are doing it *right* or *wrong*, the same way you can enjoy music as well. As long as you have the desire to create, no one can say what you make or do is not *real*.

In the upcoming chapter, we will shift our focus from discussing the industry and external factors affecting our creative confidence to time management and how we could value our creativity alongside our busy schedules. From there, we will look into perfectionism, validation, and fear.

## Notes

1 Jeff Nelson. 2021. "RuPaul's Drag Race": RuPaul explains his famous saying 'We're all born naked and the rest is drag'". Cheat Sheet. Accessed 15 January 2024. www.cheatsheet.com/entertainment/rupauls-drag-race-rupaul-explains-famous-saying-born-naked-rest-drag.html/.
2 Debbie Rohwer and Mark Rohwer. 2012. "How participants envision community music in Welsh men's choirs". Accessed 15 January 2024. https://eric.ed.gov/?id=EJ996056.
3 Ewa A. Miendlarzewska and Wiebke J. Trost. 2013. "How musical training affects cognitive development: Rhythm, reward and other modulating variables". *Frontiers in Neuroscience*, 7(279). www.ncbi.nlm.nih.gov/pmc/articles/PMC3957486/.

## Additional Resources

- Bayles, D. and T. Orland. 2001. *Art & Fear: Observations on the Perils (and Rewards) of Artmaking.* Image Continuum Press.
- Congdon, L. 2019. *Find Your Artistic Voice: The Essential Guide to Working Your Creative Magic.* Chronicle Books.
- Kleon, A. 2012. *Steal Like an Artist: 10 Things Nobody Told You About Being Creative.* Workman Publishing Company.

# No Time to Learn and Create

# 3.8

## Keywords

- procrastination
- self-belief
- career fear

## When We Do Not Have the Time to Create

We all have responsibilities that consume most of our time and energy. You might have a family and children, someone to care for or a career that keeps you busy around the clock. It is tough to imagine that there could be any time for any extra activities in our daily routines, especially those that concern self-development, creativity, or playful enjoyment. Journalist Kate Murphy summarises this issue of a modern human, in just a couple of sentences:

> Ask people at a social gathering how they are and the stock answer is *super busy*, *crazy busy* or *insanely busy*. Nobody is just *fine* anymore.[1]

It's important to acknowledge that for some, simply surviving can consume all their time and energy, leaving little room for pursuing their passions and goals. However, for those fortunate enough to have a moment to reflect, it can be valuable to examine whether we are prioritising our daily activities based on familiar routines that make us feel secure and comfortable rather than challenging ourselves to achieve personal growth. For example, assessing why you may feel guilty about using your spare time to explore your creativity or develop new skills.

Our approach to managing life is closely tied to our perception of time. How we view time impacts prioritising tasks and allocating resources to

DOI: 10.4324/9781003194484-12

achieve our goals.² Some people might find that they do not have enough time as they have so much to do. For others, one task can be overwhelming and can take a long time, whereas some are habitual procrastinators.

> ### PSYCHOLOGIST COMMENT
>
> An extract from an interview with Tricia Greenwood.³
>
> Procrastination stems from the limiting self-beliefs and thoughts that fuel our resistance to getting started. Often, we do not know why it is so difficult for us to get going, and until we do, we remain stuck.
>
> A good starting point is to recognise the stumbling mind blocks that keep us locked in our procrastinated state. Writing down our beliefs and thoughts as to why we stop ourselves from starting can focus our awareness on seeing what are the specific thoughts that are obstacles to action. Once we have identified them, we can then challenge and counter them with positive statements. For example:
>
> **Limiting beliefs and thoughts**
> I do not have the skills to succeed so I will not bother trying.
>
> **Positive statements**
> When I put my mind to it, I can succeed. I have achieved lots of things in the past. If I try more often, I am more likely to develop more skills to achieve my goals.
>
> In this way, we begin to dispute the negative thinking that encourages our procrastination, and in turn, loosen the grip it has on our behaviour.

In contrast, some people might not struggle with procrastination but have the driving power to get lots done in a short time. For outsiders, this level of productivity might look like the goal, but inside, the person can struggle with perfectionism, validation, anxiety, stress, and even burnout. So, however you might struggle with time, remember there is no perfect way to do it. Balance can look different to each person, depending on your values and goals, but recognising the insecurities that either limit or drive us can only take us towards time management that is healthy for us.

## Author's Experience

### Musician Who Does Not Prioritise Time for Music

While writing this book, I needed to evaluate my insecurities in the process. I realised I had an unhealthy relationship with work for a while. I have always prided myself as a very productive person, and if I get a project idea, I will do it with 110% effort. But with time, I have realised that this pride I feel when explaining how I have just edited 15 hours straight is not healthy or giving me the results I aim for. The workaholic mentality, which made me feel successful and valued, stems from the need to prove myself, validation, and perfectionism. I noticed how I have focused on quality over quantity, while dismissing my artistry completely, explaining it with a lack of time.

Only a couple of weeks ago, I was preparing for a regular Monday. My schedule was typical for a self-employed person: I had some admin and emails, a couple of meetings, and writing this book. That Monday morning, I chatted with my husband over breakfast about what we would be working on that day, and he asked about the song I had composed over the weekend. You see, I had forgotten that a couple of days before, I came up with a song idea that I felt very inspired about. And for the whole weekend, I did not stop singing and talking about it to my husband. But when Monday came, suddenly I had pushed away this inspiration and put my *professional* hat on. He remembered how much I loved the beat that I had made, but somehow I did not. Almost as though playtime was over, and now it was time to do some *real* work.

The most bizarre thing about this situation is that music is my job and the core of how I make my living. So, why do I, even as a professional, think music, creativity, and inspiration are not priorities that deserve a place in my daily schedule? This is why I work towards understanding the insecurities that might affect my drive and finding a better balance between work, income, and how I value my creativity.

## Why Do We Struggle to Invest Time and Money Into Arts?

When reading the news about constant cuts in arts education or listening to the older generations questioning the value of a music or art degree, it does not surprise me that we have these insecurities. When the society we live in does not give value to arts and encourage creativity, it is no wonder we struggle to provide it with the importance it deserves.

It can also feel selfish to think we deserve to give creativity time in our daily life, even if we do not want to make it a career. The guilt can come from somewhere deep down where we see creative industries as self-indulgent. Also called *career fear*, commonly seen in women, although anyone can struggle with it. Career fear is an experience that a person who suffers from low self-esteem and imposter syndrome can experience. They do not believe in their achievements or that they deserve success in their goals.[4] Therefore, when time is limited, and life gets complicated, it becomes easier to focus on others' needs instead of taking care of yourself first.

Similarly, when stress takes over, survival mode kicks in, and we feel like we are wearing blinders. Stress affects not only our mental health but also our physical functions. For example, it can increase depression and insomnia, weaken your immune system, make you miss your period, and increase headaches.[5] That is when taking care of yourself, your health, and your passions become the last things on your list to focus on.

Even though you might feel that you will never see a day when you can find time for creativity, there are ways for us to take small steps towards a healthy balance of it all. In an interview with Ski Oakenfull, an artist, keyboard player, producer, remixer, and composer, he mentioned about creativity after having a child and how it challenged him in a positive way to approach his workflows in a different way:

> When I first had a child, I was a full-time musician and very indulgent with my time. And then, as soon as I had a child, I wondered if I would have no time and how I would cope. But I found that it had the opposite effect. It just made me really focused. And anytime I had, and however exhausted I was, I would utilise that time. So it was a good example of how I integrated my family life into my musical career. Also, it made me put myself in different environments, which I also love. I would grab 10 minutes in a cafe with a laptop and do an edit of a track, or you could design a sound or do whatever. And I love just being able to utilise any moments I might have. You might do something very differently if you sit in a cafe or a big studio. I think it gives immense creative freedom to have restrictions like that.[6]

For some, the idea of a healthy and rewarding life balance can seem very far away. In these situations, seeking guidance from a professional therapist or psychologist can be helpful. If you struggle with time, but feel ready to take steps towards making more time for yourself, look at the tips at the end of this chapter, and also refer to Chapter 4, where we will discuss more about time management and how you can plan your way to your goals.

> ## OVERCOMING PROCRASTINATION AND TIME MANAGEMENT OBSTACLES – TIPS FOR MUSIC PRODUCERS
>
> 1. **Time Blocking**
>    Time Blocking is a common productivity technique that divides your time into realistic blocks. This will help you to take tasks step by step, manage expectations, and control stress. First, schedule all necessary appointments, work and other commitments, then plan a dedicated time for your learning and creative practice. It can surprise you how much *1 hour* of assigned music production time can change your life. More info about time blocking can be found from Chapter 4.
>
> 2. **Support Network and Accountability**
>    Accountability by someone you trust can increase your motivation to work towards your goals. Set monthly or weekly goals with your accountability partner, and establish a reward system for accomplishing your goals.
>    In my Patreon community, I arrange monthly challenges for musicians. It is incredible how much me and my community members have gained from this supportive community and scheduled deadlines. Find details of this community from the extra resources of this chapter.
>
> 3. **Health Comes First: Taking Care of Yourself Is an Act Towards Your Goals**
>    Remember to be patient with your progress and goals. If you are not in good health to go forward with your dreams, take care of yourself first. Everything else can wait.

## Notes

1 Kate Murphy. (25 July) 2014. "No time to think". *The New York Times*. Sunday Review.
2 Marty Nemko. (9 April) 2019. "Time management and procrastination". *Psychology Today*, n.p.

3 Tricia Greenwood. (October) 2021. Interviewed by Liina Turtonen.
4 Lauren Geall. (12 August) 2020. "Imposter Syndrome: How to stop 'career fear' holding you back". *Stylist*, n.p.
5 Ann Pietrangelo. (29 March) 2020. "The effects of stress on your body". *Healthline*, n.p.
6 Ski Oakenfull. (July) 2022. Interviewed by Liina Turtonen.

## Additional Resources

- Duhigg, C. *The Power of Habit: Why We Do What We Do in Life and Business*. Random House Publishing Group, 2012.
- Soojung-Kim Pang, A. *Rest: Why You Get More Done When You Work Less*. Basic Books, 2016.
- The School of Life. 2019. "Taking it one day at a time". YouTube. www.youtube.com/watch?v=UhWFddWz1Nk.

# I Feel the Pressure of Being Perfect

# 3.9

## Keywords

- perfectionism
- jealousy
- validation
- fear
- failure
- shame

Many of us wish we could make perfect music with the perfect workflows and gain the perfect sales stats for our music afterwards. Perfectionism is something we all experience, especially as creative people. It forces us to face our fears in terms of our art, dreams, and hopes. In parts, it can make us insecure, procrastinate, or stop going forwards with our goals, but it can also encourage and challenge us to reach new spheres.

This chapter focuses on understanding perfectionism and why it has such control over our creative confidence. It will also be an excellent start for a more profound psychological discussion about some of the most significant insecurities we have as creatives; jealousy, validation, fear, and failure. All of these subjects will be discussed separately in the following four chapters.

> Musician and Creativity Empowerment Coach Sara Belle comments:
>
> Perfection is a reaction that arises from a need to feel safe. To be safe from critique, shame or failure. It is also rooted in the idea that one's achievements equivocate to being a good, successful, perfect person.

DOI: 10.4324/9781003194484-13

> Maybe you feel that your music has to be perfect because you are worried about how you look. Perhaps you feel that your productions have to be perfect because you don't like how your voice sounds. Maybe you have to be perfect because you feel like an imposter.
> 
> To tackle perfectionism, it serves to get clear on what we are trying to protect ourselves from. Is it a fear of failure? Is it a fear of not being liked? Is it a belief that we are our work? Perhaps it's a past experience that we need a little self-compassion and forgiveness to facilitate healing? Once we understand the root cause, we can begin to dismantle the stories that we have built up around our supposed imperfectness.
> 
> Instead of aiming to be perfect, take time to play, experiment and detach from results. Remember, this should be fun! Focus on what feels good about your practice and how you can incorporate more of this. For example, if you worry about performing, do not fixate on the perfect gig, instead think about creating a shared, unique experience with your supporters.
> 
> Your creativity is a journey. To be perfect today means nothing much is happening tomorrow!! Embrace that you are constantly evolving and that there is no space for perfection when you have a growth mindset.[1]

## How Can Perfectionism Affect My Confidence, Both Good and Bad?

In music production, perfectionism shows in multiple occasions: learning, creation process, and career goals. For example, one of the biggest frustrations with my beginner students is not managing to create what they envisioned straight away. In other art forms, such as fine art, we all know what unfinished work looks like and understand what type of masterpieces the experienced artist can do. But in music production, all of this is even more confusing. Most of the songs we hear on radio and streaming services are of excellent quality. We are rarely exposed to raw, unfinished, flawed works from peers and professionals.

At the same time, if a beginner is put in front of an incomplete and mastered track, for them it could be almost impossible to pinpoint the differences. This is because active listening and understanding what makes tracks professional sounding is about years of ear training and understanding concepts like frequencies and dynamics. Alongside this, the standards of the industry and the external pressures of this career are endless. No wonder students entering the industry, or even more seasoned producers, might experience consuming perfectionism, chasing up the perfect sounds, the perfect statistics, and perfect music.

Perfectionism is all about how we manage our expectations. In our imagination, we picture a perfect outcome or result for our goals. But in adulthood, there are no rewards or gold star for our success. We keep reaching for objectives, looking for perfection, which in reality does not exist. And we must face the fact that sometimes our grand visions of the future do not match the modest reality of life.[2]

But try not to feel disheartened or think there is no point in aiming high, as goals might seem impossible to achieve. We can strive for excellence in our plans and find balance in our perfectionism. We can understand that there are different ways to be a perfectionist and recognise the harmful patterns they can cause us, while harnessing the positive qualities that push us forward. Aiming for balance and not for *perfect*.

Psychologists Paul Hewitt and Gordon Flett have recognised how perfectionism is a multidimensional spectrum with three main varieties:[3]

- **Self-oriented perfectionists** have high standards for themselves, especially in their careers and personal lives. They are motivated to go after their goals, be organised, and have high success rates.
- **Other-oriented perfectionists** can hold others countable and expect perfection from family members, colleagues, and other people they might connect strongly with.
- **Socially prescribed perfectionists** can feel highly self-critical. They can experience pressure from the people around them to do and be the best in their practice, creating anxiety and difficulties with confidence.

Perfectionism can manifest in many ways, and it's likely we all have some degree of these traits. For some, it is the fear that making a wrong move will have disappointing consequences, leading to anxiety around decision-making. For others, it is a need for control over every aspect of their lives, which can be both motivating and limiting. In music, it is common to push ourselves too hard to achieve perfection in order to gain the approval of others, leading to creative blocks, exhaustion, and burnout.

Perfectionism can make us push ourselves for more remarkable achievements or improve our work quality. But even if perfectionism seemingly enhances our performance, it does not mean it does not affect our mental and physical health. That is why we need to recognise how it affects our creative practices and how we can find a balanced relationship with our perfectionist tendencies.

My experience with perfectionism has forced me to confront my insecurities. When I started my YouTube channel, I was determined to grow it and heard that posting consistently was the key to success. As a perfectionist, I took this advice too seriously, and for over three years, I posted every Sunday at 8 pm without fail. I was afraid that if I stopped, even for a week, I would be a failure and lose all the progress I had made. Though I am proud of the discipline that helped me build my career, it eventually led to burnout. I realised that working this way was not sustainable and was affecting both me and my content. Ironically, after I stopped, I started gaining more monthly followers, and my videos did better than when I posted once a week. Now, my followers say they like my content more because I seem happier and enjoy it more.

Through my experience, I have learned that kindness towards yourself is the key to finding balance with perfectionism. When we feel like the results of our efforts are insufficient or we get frustrated that everything we try never seems to work, forcing ourselves to work harder is not the solution. Instead, we need to be kind and remind ourselves that our negative thoughts are not valid and that slowing down is okay. We should strive to find motivation through excitement and enjoyment without constantly focusing on improvement or fearing failure.

---

Some thoughts that can help you overcome your perfectionism:

- Remember, there is no failure in creativity.
- We cannot control what creative ideas come through us at any given moment. In a way, we are not responsible of our creativity and therefore, we should not try to limit, filter, or control it. Let whatever flows through you come out and decide what you wish to do with it later. These are called *Divergent* and *Converged* thinking models, which we will discuss further in Chapter 4.
- Your perspective of what you have created is clouded by your fears of failure and your expectations. Start to imagine that the song you were working on was created by someone else and distance your judgement from it. Look at it with a new perspective and trust the process.
- Be kind to yourself and your ideas. Would you talk to your friend about their art like you speak to yourself?
- Leave some time and come back later. Maybe at the moment of creating the song, you needed to make it perfect, making you feel like it was never going to be good enough. But time gives us perspective. Come back to it with new ears, and try to look at it with kindness. See its potential and not the weaknesses.

## Navigating Our Goals

Finding balance in perfectionism comes from matching our time and effort to our expectations. Problems come when we do not understand the whole reality of what it takes to achieve the set goals. The issue, therefore, is not about aiming too high, but learning what your perfect vision demands (refer to Figure 3.7).[4]

For example, if you wish to become a successful musician, you need to study and understand how it can be realistically achieved and in what time frame. This vision would need plenty of patience, dedication, and time. Besides learning the music business, running a company, and the social media and marketing side of artistry, finding pleasure in the process is crucial. While striving for greatness with a sense of purpose is a positive aspect of perfectionism, it is important to be aware of the realistic steps you can take towards your goals in order to avoid feeling overwhelmed and discouraged.

When I first began pursuing my music and YouTube careers in my twenties, I aspired to build a sustainable career. However, I did not fully understand what that meant then, as I felt that gaining global fame and recognition was the only way to achieve a career as an artist. Looking back, I realise that these dreams were naive and influenced by the image that the music industry portrays: either you are a struggling artist or a superstar. I did not realise that there could be something in between.

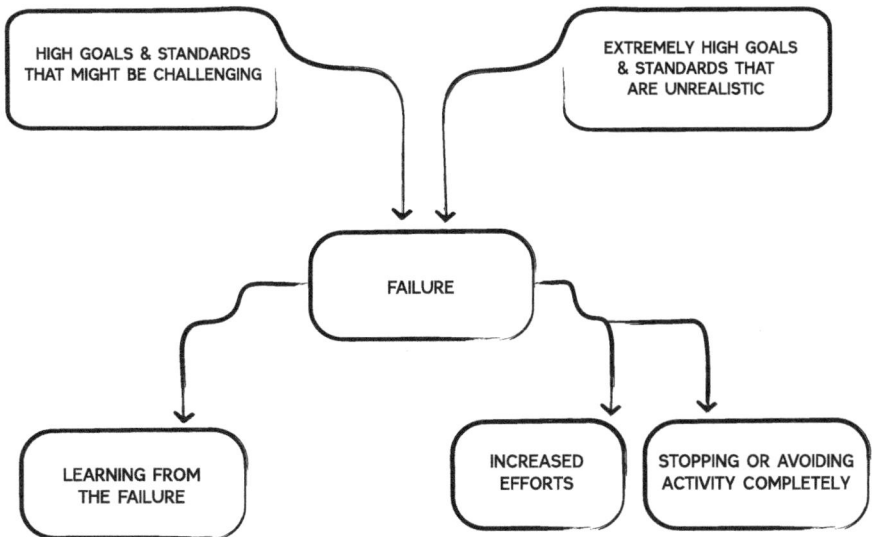

FIGURE 3.7 The Pattern of Perfectionism by Jessica Rohlfing Pyror

Illustrated by Emma Holdway

Only later did I discover that artists can have a fulfilling and sustainable career in the arts without the need for viral videos or massive fame. This realisation was like a weight lifted off my shoulders. I started following artists who were a little older than me and had successful full-time music careers in their niche. It showed me that this was what I wanted – not huge fame, but a loving community allowing me to make a living doing what I love.

Once I clarified this goal, redefined what success meant for me, and saw that it was all achievable, my perfectionism started to slowly fly out of the window. I no longer needed to prove myself to imaginary people who I thought were gatekeeping what I should make and standing in the way of my success. Instead, I started making music that I truly loved and began to believe that there would be people out there who would love it just as it is.

To address the negative impact that perfectionism can have on creative work, we must educate ourselves on how others have achieved their goals and overcome obstacles in their journeys. By learning about the real struggles and efforts of others, you can better understand the practical amount of effort and time required to achieve your own goals. That is why Chapter 5 of this book features stories from industry professionals who share their insecurities and struggles, showing us that we are not alone. These successful people have also dealt with their obstacles, and by learning from their experiences, we can better equip ourselves to overcome our challenges.

When we feel overwhelmed, consumed, or controlled by our perfectionism, the first step to changing it is to recognise our thinking patterns. We can give alternative thinking patterns for ourselves while separating which thoughts are harmful and which can positively influence us.[5] This can then help you weigh your goal and break down the realistic approach to achieve it, motivating you to move forward and not be stopped by the fear of failure.

## Examples of Alternative Thinking Patterns for Perfectionist Music Producers

These are examples from the author's life. You can use a similar structure to create dialogue for the moments you need a reminder of the impacts of your perfectionism:

- **1st Statement:** I cannot finish this song, as it does not sound like what I wanted.

- **Answer:** By finishing this song, I might experience the feeling of success, even if the track is not exactly what I wanted it to be. I could practice finishing songs, even if they are not always masterpieces. The more I finish my music, the more likely I am to make something I love. After they are done, I can decide what to do with them.
- **2nd Statement:** I want to publish a song every month to get my Spotify algorithm higher. I am not sure if I will have the time to do it, but as I promised myself, I will do it. I need to complete this task. Otherwise, what is the point of publishing music if you do not give it everything you got?
- **Answer**: I recognise that I want to publish more music regularly. It is excellent that I have the passion for making it and being able to monetise it. But pushing myself to do it regularly, when other commitments in my life also need my attention, is currently too much to ask. Maybe a better alternative is aim to make music monthly, without the goal of publishing it. I will schedule four hours every month just for making music and nothing else. If something more comes out of that, well, that is excellent, but it takes away the pressure to succeed in something that might not be realistic right now.

In Chapter 4, you will learn further about navigating and organising your goals in music and finding ways to overcome the most significant creative blocks perfectionism might cause you.

## Notes

1. Sara Belle. (October) 2021. Interviewed by Liina Turtonen.
2. The School of Life. 2017. "The problem with perfectionism". YouTube. Accessed 15 January 2024. https://youtu.be/g8pti-Swh_E.
3. John Gregory Boyle, Donald H. Saklofske, and Gerald Matthews, eds. 2015. *Measures of Personality and Social Psychological Constructs*. Elsevier/Academic Press.
4. Northwestern University. (12 April) 2019. "Pushing back on perfectionism: How to be happily imperfect". Accessed 15 January 2024. https://counseling.northwestern.edu/blog/maladaptive-perfectionism-coping-strategies/.
5. Martin M. Anthony. Department of Psychology, Ryerson University. (9 April) 2015. Accessed 15 January 2024. https://adaa.org/sites/default/files/Antony_MasterClinician.pdf.

## Additional Resources

- Luyken, C. 2017. *The Book of Mistakes*. Dial Books.
- Struthless. 2020. "Advice for perfectionists & procrastinators: The 70% rule". YouTube. https://youtu.be/SxA69uUGEUI.
- Saltzberg, B. 2010. *Beautiful Oops!* Chronicle Books.
- The School of Life. 2019. "The perfectionist trap". YouTube. https://youtu.be/BY6bGhcnDDs.

# I Feel Jealousy and Envy

# 3.10

## Keyword

- envy

## Jealousy, Envy, and Art

It is impossible not to compare ourselves to others in the arts. The modern music industry is built on measuring our value with streams, likes, followers and retention minutes. It can become highly disheartening to learn music skills for years, then create and perfect your craft for it to be published and instantly judged by people and the algorithm. Ultimately, this will form a sense of failure and naturally inferior complexities towards peers, feeling like the music industry does not have enough space for all of us. The circles seem small, and you need to know the right people to get anywhere in your career. It is common to rank people higher than us, imagining they might be more confident, luckier, or more talented to deserve their place. That is why we often forget that all of these people started from somewhere.

Due to the vast number of artists publishing music daily, the competition for visibility is hard. To get streams as a self-publishing artist, competing with musicians with representation or labels backing them up can be challenging. And even if you have a record deal, you need to work hard to get the support some of the other artists on the same label might get. In whatever stage of your career you might be at, the industry is focusing on competition and success. As individuals, this can naturally cause feelings of jealousy and envy.

Jealousy comes from suspicion of rivalry. Perhaps you feel that so many other artists are trying to get noticed, and everyone at the same level as you is competition. If they get an opportunity, it might feel unfair, and if others get to the goals before you, it can feel like it is getting away from you. Alternatively, envy is longing for something that others might have, and you feel lacking. You can feel envious of someone having a record deal and wonder why you do not have it, even though your music is good or even better than theirs.

In arts, especially in the era of social media, we often only see the success of others without the failure in their careers and journeys. This can distort the idea of reality and make peers seem more significant and better than what we feel in ourselves, making our losses feel like they only happen to us and not to anyone else. The feeling of failure can trigger intense sensations of jealousy and envy. Just remember, it is normal to experience these feelings.

We can feel jealousy and envy when we feel threatened, afraid of losing what we already have or second-guessing qualities we believe we have. They can both distort your thinking, raising feelings of fear, anxiety, anger, sadness, shame, and loss.[1] Similarly, as with perfectionism, jealousy and envy are the disconnections between our imagination and evidence of the events actually happening at the time.

## Causes of Jealousy and Envy[2]
(Refer to Figure 3.8.)

- **Mind-reading** is assuming what people think and believe. Mind-reading is extremely common in the audio industry, where technological knowledge can be intimidating to approach. The community encourages knowledge hierarchy, meaning the more you know, the better your skills are. Therefore, we often assume others know more, making us question our abilities to learn and use the different technological tools.
- **Personalising** is when we interpret everything happening to be related to ourselves. For example, perhaps you send your song to be featured in your favourite music blog, and they answer that it does not fit the genre they are currently looking for. Instead of accepting that the rejection was about the genre and timing, you might still believe they said no because they did not like the track.
- **Fortune-telling** happens when our actions are impacted by what others do around us. We might start to compare our potential and skills to others, predicting we will be less competent at the task, possibly preventing us from trying it out in the first place. Maybe you feel like your friends will get a festival gig instead of you, so you will not even apply.

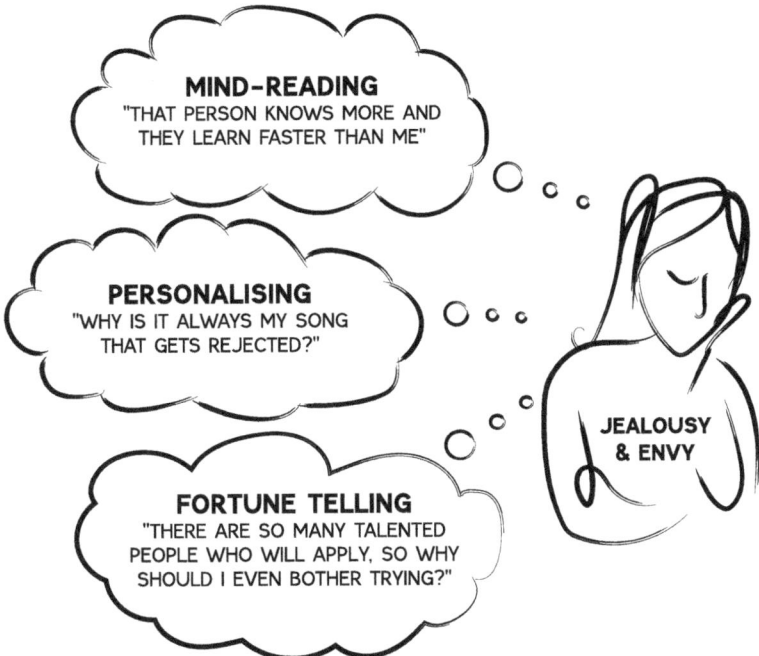

FIGURE 3.8 Illustrated by Emma Holdway

Musician and Creativity Empowerment Coach Sara Belle comments:

> We encounter jealousy and envy when we feel that someone has what we should have or they are doing things that we wish we were doing. Sometimes these emotions arise because we are worried that we are being pushed out of things that we should be included in.
>
> Despite feeling uncomfortable with these emotions, we can actually learn a lot from them as they inform us of things we want to manifest in our own creative practices. Maybe seeing your peers on tour sparks up a little green flame in the pit of your stomach. Great, look into that! Sounds like you are up for a tour! So, get thinking about what things you need to put in place, create a plan, and sketch out a manageable timeline and get going!
>
> Another great approach is to connect with people!! Jealousy and envy occur in situations that we feel we should be part of, belong to, be equal in. If you are looking at someone's mega polished music video and you are feeling the pinch of envy, reach

> out to them and ask for some advice, you would be surprised how many people want to help. You then get to transmute the experience of jealousy and envy, which can be very isolating, into feeling part of a community and feeling supported.
>
> Finally, remember that we all have different starting points, challenges, advantages, and lessons to overcome in our journeys, so there can never be a direct comparison with where you are and where others are. Know that what you see of someone's achievements is the result of many little steps. Be compassionate to yourself when you take inspiration from the achievements of others and be comfortable forging your own path at your own speed.[3]

## Jealousy, Envy, and Music Production

From a young age, I have felt both jealousy and envy often. Whenever I saw someone in a position I wished to be, it felt like a physical painful twist sensation in my chest. In parts, I think jealousy and envy have both stopped me from reaching my full potential and pushing me forward. For example, when I released music and sent it to radios and media without response, it felt so disheartening that it made me question my skills as a musician. It can be exhausting to try endless times when others seem to get to the goals faster. On the other hand, seeing others reaching the goals you dream about while feeling the painful envy in my chest has made me work harder and stopped me from giving up when trying has gotten exhausting.

By pushing towards my goals and not giving up, I have noticed how envy and jealousy have become easier to deal with. When you feel more secure in your skills and achievements, it is more manageable to accept your position in the industry and realise that there is room for everyone. Although I am not saying these feelings will ever entirely disappear, we can practise so that jealousy and envy are not something that consumes our lives but takes us forward.

In music production, we find that these feelings appear in several situations. Sometimes, it is easier to be aware of them, and sometimes they exist without clear indication. For example, when we see a picture of a peer hired to produce a song in a big studio for a prolific musician, it is easy to recognise envy or jealousy. But when we are in the middle of a song-making process, we might get creative blocks related to these feelings without a clear indication of where they are coming from.

In Chapter 4, we will discuss plans and creative workflows. These are designed to eliminate the chances of feeling jealousy and envy in the creative process, even when experiencing feelings you might not be aware of at that moment.

## Turn Jealousy and Envy Into Your Strength with This Writing Exercise
(Refer to Figure 3.9.)

1. **Start by finding awareness in your jealousy and envy.** But first, do remember not to be frightened of these feelings, as they are typical to all of us. Maybe you find yourself listening to someone else's music, and you start to feel like everything you do is worse, or your chances of hearing the great ideas in your music are lower. Write down the question: What do I feel? And start listing the feelings these thoughts bring in your mind and body. Maybe you feel like your chest gets heavier, your hands sweat, or you find it hard to breathe. Write down what you experience, and let yourself experience them without shutting them down. By experiencing them and being aware of how and why we feel, we can better remember that some of our feelings might be mind-reading, personalising, and fortune-telling, where there is a separation between the evidence and what we imagine.
2. **Now write down a second question: What is my journey, and what is theirs?** Everyone has their struggles, obstacles, and journeys towards their goals. Find a way to focus on your story and how you become the person you are today. Similarly, try to see through the whole picture of the people you might idolise or compare yourself to. What is their story, and wonder what obstacles they needed to overcome to be in their current position? Perspective and sympathy towards yourself and others can lower the gap between being consumed by jealousy and envy and turning it into a strength.
3. **Finally, the third question: What evidence do I have of my strengths and achievements?** Write down the things you are proud of, the moments you achieved specific goals, and the things that brought you the most joy. Remember to look at your success through enjoyment rather than what others might give higher value. By looking at these achievements, consider them as facts, manifesting who you are and who you can become.

Repeat this exercise every time you feel jealousy and envy are taking a negative hold on you, and you wish to find a belief in your journey and strengths once again. In the next chapter we will discuss validation, a related feeling to jealousy and envy, and how it affects our creative processes.

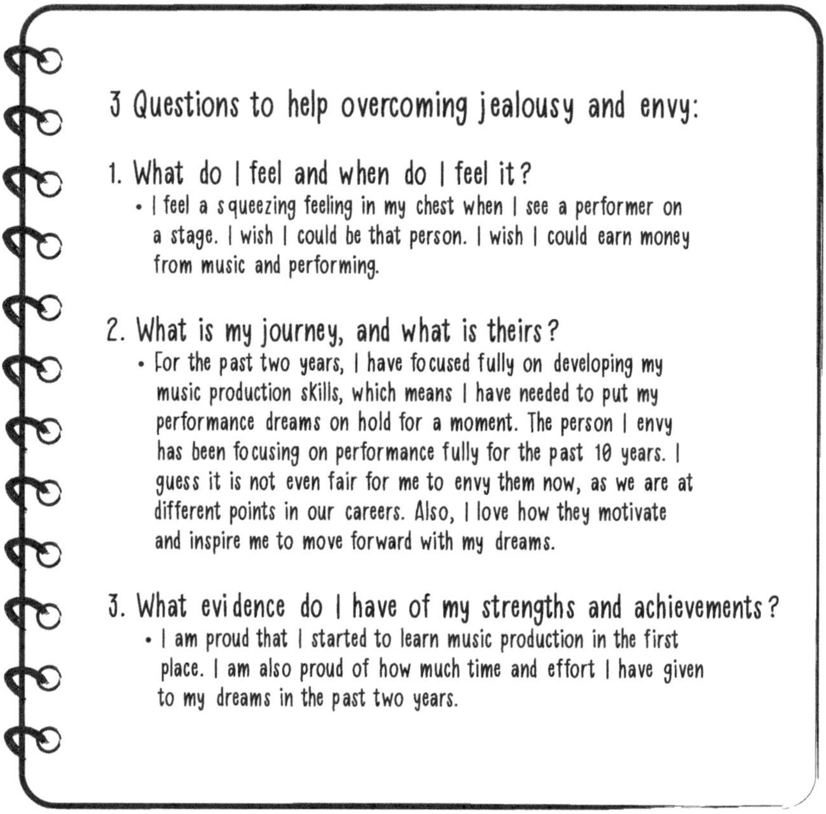

> **3 Questions to help overcoming jealousy and envy:**
>
> 1. **What do I feel and when do I feel it?**
>    - I feel a squeezing feeling in my chest when I see a performer on a stage. I wish I could be that person. I wish I could earn money from music and performing.
>
> 2. **What is my journey, and what is theirs?**
>    - For the past two years, I have focused fully on developing my music production skills, which means I have needed to put my performance dreams on hold for a moment. The person I envy has been focusing on performance fully for the past 10 years. I guess it is not even fair for me to envy them now, as we are at different points in our careers. Also, I love how they motivate and inspire me to move forward with my dreams.
>
> 3. **What evidence do I have of my strengths and achievements?**
>    - I am proud that I started to learn music production in the first place. I am also proud of how much time and effort I have given to my dreams in the past two years.

FIGURE 3.9 Illustrated by Emma Holdway

## Notes

1. Susan M. Pfeiffer and Paul T. P. Wong. 1989. "Multidimensional jealousy." *Journal of Social and Personal Relationships*, 6(2): 181–196. https://doi.org/10.1177/026540758900600203.
2. BrainCraft. 2018. "How jealousy distorts your thinking". YouTube. Accessed 15 January 2024. https://youtu.be/Joy6yZs47Lk.
3. Sara Belle. (October) 2021. Interviewed by Liina Turtonen.

## Additional Resources

- Brown, B. 2010. *The Gifts of Imperfection: Let Go of Who You Think You're Supposed to Be and Embrace Who You Are.* Hazelden Publishing.
- DeAngeli, C. 2016. *Life Without Envy: Ego Management for Creative People.* Chronicle Books.

# I Cannot Stop Asking for Validation    3.11

## Keyword

- self-regulation (also emotional regulation)

Throughout this book, we have talked about overcoming the challenges that may come when trying to express our creativity authentically in the music and audio industries. This chapter will dive deeper into the root cause of these difficulties – the need for validation.

## What Is Validation?

We all want to be heard, noticed, and belong. We seek validation for our thoughts, feelings, and accomplishments, which is entirely normal and healthy. However, this need for validation is further amplified in the creative industry since the art created by artists is meant to be seen and heard by others. Seeking approval and validation as an artist is not wrong; in fact, it is a basic human need that we all require in one way or another.

    Think of a moment when an important person to you listens to your latest song. Even if they do not particularly enjoy it, simply acknowledging the effort and dedication you put into creating it can mean the world to you. As much as validation can make us strong and feel powerful, it can also feel painful and increase our anxiety if we look at it from the wrong places. Therefore, validation can either pull us back from our full creative potential or allow us to build confidence and find strength when navigating insecurities.

DOI: 10.4324/9781003194484-15

As a musician, you might feel frightened to publish new music or struggle to figure out what type of music you wish to make. For these emotional battles, it is common to hear answers such as *just do it for yourself* and *do not think what others think*. But in reality, it is impossible not to be affected by the modern music industry around us, and fully feel like you can create with complete honesty.

We often get scared or feel shame about the things we value the most. For example, you might be creating a song, and in the middle of the process, you get an idea for lyrics. The words are very personal and relate to something important in your life. But before you even manage to write them down, you say no to your idea, feeling embarrassed. What if others feel like they are cheesy, too emotional, or they expose sides of you that might make you vulnerable to judgement?

As much as we wish to be fully authentically ourselves, speak our truth in our lyrics and write only music that we feel passionate about, we sensor, sometimes unconsciously, what we put out because of fear. This means we hide part of who we are so we do not get left out and fail to blend in.

Therefore, the need for validation can confuse us about our identity, and we might struggle to be uniquely ourselves. This can especially impact our confidence in the arts, where we may find it challenging to create music that we genuinely love, free from the influence of genres and expectations. As a result, we might strive towards goals that do not align with our natural abilities or passions, causing confusion about our aspirations and whose dreams we are even chasing. This can lead to feelings of inadequacy as we compare our perceived failures to the apparent successes of others.

> Musician and Creativity Empowerment Coach Sara Belle comments:
>
> The need for validation is something that we all share as humans, as social creatures. It's the way by which we co-create our shared reality. To validate is to agree that something is real, that it is good or that it meets a standard agreed by a community or an authority.
>
> In our creative practice it is natural to look for validation. However, dependence on receiving validation for our creative work is like trying to drive a car from the outside – it is extremely difficult to make progress and it certainly is not comfortable!
>
> A need for validation is often representative of feeling low on self-trust, self-belief, and self-efficacy. We start creating work that we think others will like, often to the detriment of our authenticity.

> To get back in the driver's seat of our creativity, we have to tune into our self-trust; that deep inner knowing that serves to guide us once we create a space to hear it. Start by taking a moment to think about your "why" – "why are you creating this work?".
>
> Our "why" is our foundation, born out of a need to address questions or experiences unique to us. It exists independent from external validation. It's the passionate fire within our soul that fuels our creativity.
>
> When you feel that you cannot trust that your work is real, that it is good, that it fulfils a standard, revisit your "why". Is this work true to your "why"? In doing so, we learn to tune into our self-trust and build up our own ways by which we can understand and value our creative power.[1]

## Emotional Bypassing

For most of us, talking about emotions can feel inconvenient and exhausting in everyday moments. Often we feel like talking about feelings can make others and ourselves uncomfortable. It can also be confusing and overwhelming to try to navigate precisely what it is that we are thinking. That is why we often push our feelings away and do the same to children.

*Emotional bypassing* is when the people around us do not see or listen to why we feel something but either ignore or go around the conversation with a simple result.[2] For example, let's say that you are at a garden party and chatting with your friends about your new song. You feel slightly upset as it has not done as well as you hoped. Your friends might feel uncomfortable with this statement and try to make you feel better by saying, "Do not worry about it now. Let's enjoy the sun and stop thinking about the stats".

Even if your friends just meant to help you, encouraging you to forget it at that moment might distract you from the negative feelings but it does not take them away. Instead, if your friends would have recognised that this has upset you by saying, "I can see that this upsets you. Do you want to chat about it?" This emotional recognition can immediately make you feel better about it.

Emotional bypassing is something we learn as a child and carry to our adulthood, and pushing away others' feelings is something many of us might not even notice that we are doing. If you struggle with this now, it might be that there were not many people around you as a child who would fully listen to how you were feeling. When we do not feel seen and noticed

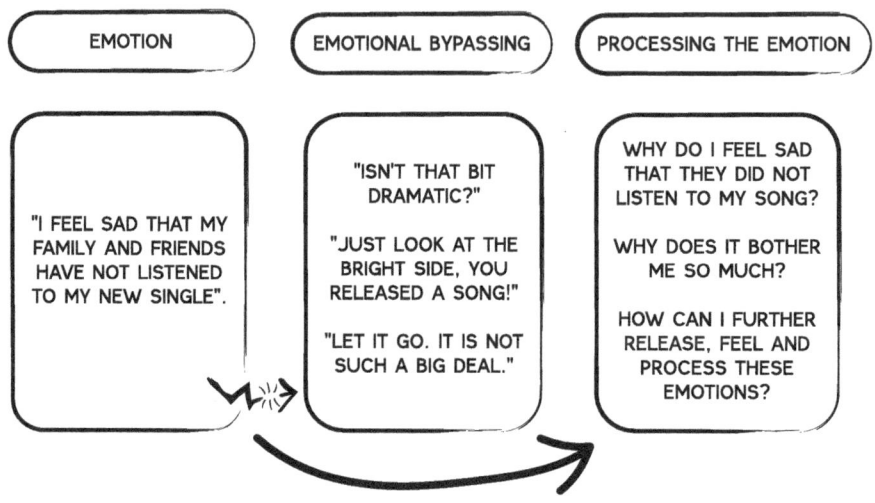

FIGURE 3.10 Emotional bypassing
Illustrated by Emma Holdway

as a child, we often take the learned behaviour models to form childhood to adulthood, relationships, and children.

Experiencing emotional bypassing as a child can lead to intense sensations of need for validation. As adults, we can learn to recognise when we do it to others and see when it is done to us. This means when we notice someone is bypassing our feelings, we can mention it directly to the person. For example, we can ask if the person would be generous enough to listen to your thoughts for a moment, regardless of how big or small they might seem. The same applies to our music and art.

If you feel like someone close to you is ignoring what you have created, you can openly ask them to listen to it as it means a lot to you. Most of the time, the person might have just not realised that listening to your music is so important to you, whereas sometimes, even if we ask for someone's validation, we might not get the response we expected. It can hurt to realise that there are people around us who will never notice our efforts or achievements the way we wish them to. This can make you sad and disappointed, which is entirely understandable. At the end of this chapter, there are some pointers on how you can find techniques to overcome these feelings.

## Where and How Do You Get Your Validation?

The need to have our emotions validated is to do with our self-worth and how we have learned to regulate our feelings. The more we focus on others' opinions, the less we start to value our ideas. Although we will always have a choice to determine our worth, we need to understand our values and what we are genuinely passionate about. We should avoid the temptation to hide our interests to seek approval or validation from others. By embracing and being truthful about what we truly value, we can experience a sense of liberation and take pride in our own self-worth.

When we are put into situations where we might feel lack of validation it can trigger plenty of feelings in us, making us react fast, spiral in our thoughts, or feel anxiety. The inner fight to deal with these situations is called *self-regulation* or *emotional regulation*. This is a skill, that with practise allows you to be more mindful on how you choose to experience and show your feelings.[3] Even though it is healthy to feel all emotions, sometimes in situations where we as an artist can feel insecurities, the ability to self-regulate becomes extremely useful, as it helps us recognise when the feelings are unhelpful for our progress and creativity.

For example, I grew up making classical music, and in my early twenties, I mainly created music as a singer-songwriter. On many occasions, I would hear my mother comment on electronic music and how she disliked its thumping bass sounds or too busy melodies. I grew interested in EDM (electronic dance music) and especially house music with time, but somehow I always felt apprehensive about creating or publishing it. The first house track I made was probably one of my favourite songs I have ever done but showing it to my mother was nerve-racking. It was a song I was sure she might dislike, but as soon as I stopped asking for her validation of my music, I felt an enormous sense of relief. My music is determined only by what I feel, not by external assurance.

After this, I realised that people around me do not need to like my music to be able to support me in my art. Now, whenever I release music, I share what I have created by accepting that validation comes in many forms. My mother is not the biggest fan of EDM, but her support comes in the way of noticing my hard work and admiring how I have found creative input that makes me happy. Whereas from the music communities that listen to EDM, I get more music-related validation as they are people who truly like the sound of the songs. Understanding that both of these forms of validation are equally valuable helps me manage my expectations for where and how the assurance can be received.

## Creativity and Hate Comments

In this chapter, *hate* is defined by words, actions, or gestures that cross the border from honesty, to intentionally spiteful or hurtful. It is often an unnecessary comment that can take someone's confidence down without any good reason to state it in the first place.

By following our values and passions, we might have situations where people do not agree or like what we are doing. This makes us vulnerable and can be especially hard if the person disagreeing with you is someone close to you.

Especially in the music industry, we need to deal with the fact that what we create is not going to please everyone. As mentioned in this book, music has a significant connection to our identities and how we define others and ourselves as humans. This can heighten when we publish our music for the first time. The people receiving it are no longer from the supportive bubble you grew up with, but strangers with their direct take on what you have created.

This is where the lines between honesty, directness and hate often fade. Someone might feel it appropriate to comment and evaluate what they have heard with a simple "I do not like this at all. You should learn the techniques way better before publishing anything". This type of comment can be meant as feedback, but it is not constructive and can easily be very hurtful to the artist. It can make them feel attacked, whereas the listener might think that music is there for people to consume and that feedback can be given in the same way that anything can be bought and reviewed online.

They might think that "maybe if they hear the truth, they would not embarrass themselves and avoid feeling further pain". But other people's opinions are not facts. They are merely navigational tools to figure out what you wish to take and what to leave. Your ideas might make the listener uncomfortable for some reason, but it is not your job to be responsible for other people's triggers. That is why we need to learn tools to protect our confidence in a world where we cannot control how everyone approaches, uses, and talks about art.

The embarrassment, shame, sadness, and other negative feelings we might feel from these situations should not stop us from reaching our creative potential. Still, it can be treated as a reminder of how we should not create, only to be accepted. Because if our art comes from the place of enjoyment, self-expression, and passion, it will also help us to take on feedback constructively. You will only take on board the points that resonate with your values and leave the perspectives that exist only to please others.

If you are someone who has experienced hate comments, bullying, or other traumatic events affecting your mental health, you will find resources to point towards organisations that can help. Also, in Chapter 5, you can read interviews from industry professionals who address how negative and hate comments affect their confidence in the modern music industry.

## Feedback and Validation

It is important not to mix this conversation about validation and feedback with collaboration. Working together with others and having a healthy feedback loop can be a highly beneficial and positive experience for your creativity and confidence. But how you keep these interactions helpful to your development can depend on the relationship you have with validation.

Indeed, feedback is an essential tool to learn and develop. But creative confidence is knowing where you stand with your ideas and your vision, and fully believing in it. When asking for an opinion, take what you need and learn to leave the thoughts that can harm or prevent your creative workflow. If we start taking on other people's fears as facts, we can get lost in what we think is best for us and our creativity.

So, the next time you get an idea or vision, ask yourself if it is worthy of asking for feedback before you know where you stand with it. Learn to believe in your ideas and trust that there is no right or wrong in creativity.

For example, when I started my YouTube channel, I had dreamt and joked about it for years before doing it. I had a genuine passion for it and felt like it was something I needed to do. So when I had enough courage, I started doing episodes without telling anyone. I was scared that a single fear in the eyes of my loved ones could transfer it to my head. As much as they could have supported me at the beginning of my journey, I was too passionate and determined to risk it. I needed to do it first myself, and when I had landed on my feet with my confidence, the support from my friends and family felt better than ever.

## Creative Network, Accountability, and Validation

The need for validation does not only come from the people around us. With social media constantly growing, we have started gaining positive support from strangers. In my personal experience as a music producer, musician, and online content creator, I have seen how the internet can positively affect creatives.

First, it is easier to find people to identify with and gain confidence from seeing how similar people to you have achieved specific goals and defeated obstacles. Second, it pushes you to socialise with like-minded people and join moderated communities to encourage, collaborate, and create positive feedback loops. Third, finding people you feel comfortable around can create a feeling of accountability, where you wish to make more in a safe environment with people you know will only lift you up.

Do not get me wrong, the internet is also full of groups and communities that can negatively influence confidence and creativity. Still, with learning to ban all the negativity from your feed and researching these praising communities, the internet also has a positive side.

Here are some of my favourite platforms and communities to look into:

- Patreon – I especially recommend joining my LNA Does Audio Stuff community which specialises in supporting individual goals, community, accountability, and enhancing creative confidence in music-making. In this group, I have seen the power of positive encouragement and how a kind internet community can bring people together worldwide and help them all with their creative confidence.
- Discord – Several different great groups for musicians and creatives.
- Facebook – The groups I enjoy the most are Ableton Live Users and 2% Rising (for gender minorities in audio).

## My Top 5 Tips to Overcome Harmful Trades of Validation

1. **Create an alter ego artist profile.** If you struggle to finish and publish music because of the fears and insecurities validation might bring, consider creating alter ego artist channels. This is where you can post without the need for perfectionism or fearing what others will think. You will also get to make music without the pressures of advertising it or pushing it forward all the time. You can simultaneously have several artist channels on the go, one public and one for you to practise your confidence in releasing music. The important point of this project is that you will not tell anyone about it. Otherwise, the significance of it disappears. Another option is that you will do this for a limited time in secret and then go public with it after a while. Just do what is best for you. I have one of these where I have been posting music for a while now. But obviously, I will not tell you what it is called. It is between me and my creativity.
2. **Stop asking for validation from toxic places.** We all have some close people around us who we must avoid on the release day or when we have good news to share. As much as it breaks our hearts, we must accept that some people around us can never give us the validation we need. It might be that something in their past makes it hard for them to notice and be happy for others' success, or maybe they just struggle to give a compliment. Whatever the reason is, we can care for them and ourselves at the same time. One way of protecting our confidence is not

to share what we have done. It can feel difficult and sad to do, but with this, you keep the success and sense of accomplishment within yourself.
3. **Focus on people that are 100% in your corner.** Instead of looking for validation from places you know you will never get it from, focus on finding it from people who fully support whatever you do. These people do not project their fears on your dreams, they do not even question your silliest creative adventures and will see the effort behind your achievements. You might find them in your inner circle but also from a stranger. Reach out to like-minded people online and locally, and join clubs, groups, and workshops. You can even suggest people meet once a week and talk about all things music, production, and audio. Share tracks and encourage supportive, non-judgemental, kind communication. These become the people to lean on for healthy validation when needed.
4. **Learn to give yourself the love and validation you deserve.** As much as you can look for validation from others, you can also learn to provide it yourself. This means finding ways to comfort yourself when you get an overwhelming urge for validation or when the lack of it causes spirals in your confidence. As with perfectionism, jealousy, and envy, it is good to focus on giving yourself understanding and empathy. Remember your journey and how far you have come, and learn to believe in your achievements again.
5. **If you are an artist, think of social media as your job.** As modern musicians and audio professionals, social media plays a massive part in our careers. Whether raising a professional profile, promoting new music or networking, it is very hard not to feel the pressure or stop feeling the need for validation. Therefore, I have figured out one little mental trick: separate social media from the real me. Social media is work and nothing personal. But even though I share bits of my real life and some deep thoughts about myself, all of these authentic pieces of me are carefully curated to fit the purpose and brand without consuming me personally. This is how the algorithm, likes, and other stats are never about me but the business persona I choose to portray.

## Notes

1 Sara Belle. (October) 2021. Interviewed by Liina Turtonen.
2 The School of Life. 2020. "Why we need to feel heard". YouTube. Accessed 15 January 2024. https://youtu.be/hnQwaVnv-FA.

3 American Psychological Association. 2023. "Emotion regulation with James J. Gross". YouTube. Accessed 8 November 2024. https://youtu.be/Yd6hR1qCfSM?si=xj_paGuuT1WDTT-V.

## Additional Resources

- Better Help. www.betterhelp.com.
- Help Musicians. www.helpmusicians.org.uk.
- I Was Just Thinking. 2023. "Rainy day chat about validation". YouTube. https://youtu.be/iyA1GWtG9FU?si=Qd65q0_o6WqEqnw4.
- Knight, S. 2015. *The Life-Changing Magic of Not Giving a F\*\*k: A No F\*cks Given Guide.* Little, Brown and Company.
- LNA Does Audio Stuff. 2022. "How to deal with hate as a musician". YouTube. https://youtu.be/g8h17X571SA.
- Mental Health Resources. www.mentalhealthfirstaid.org/mental-health-resources/.
- Mid UK. www.mind.org.uk/information-support/guides-to-support-and-services/seeking-help-for-a-mental-health-problem/where-to-start/.
- Paulus, F. W., S. Ohmann, E. Möler, P. Plener, and C. Popow. 2021. "Emotional dysregulation in children and adolescents with psychiatric disorders: Narrative review". *Frontiers in Psychology.* www.ncbi.nlm.nih.gov/pmc/articles/PMC8573252/.
- The School of Life. 2020. "Why we need to feel heard". YouTube. https://youtu.be/hnQwaVnv-FA.

# I Fear Failure                            3.12

## Keywords

- fear
- DAW
- emotional bypassing
- failure

## How Fear Shows

Have you ever felt like you feel something so strong that you get the need to make it into a piece of music or sound? But as soon as you start putting notes to a DAW or writing words down on paper, your mind tells you to reconsider. The story you felt a moment ago now feels a bit silly, or your emotion makes you feel embarrassed. This is our fear intervening with our self-expression.

There are several reasons why fear shows itself in these moments. The reason for it showing in creative moments originates from several different external pressures, expectations, and social structures. As discussed in the last three chapters, perfectionism, jealousy, envy, and validation are a big part of our fear. But it is good to remember that everyone feels fear, even if they do not look like it. Fear is always with us, every minute of every day, even if we are not always aware of it. It comes in many shapes and forms, but it is in a threatening situation when we notice it the most.

One common cause is that we are taught to dismiss our emotions. They can make us uncomfortable and uneasy when we feel them surfacing. This is called *emotional bypassing*, where we do not let ourselves experience negative feelings at the moment or do not find ways to acknowledge and process them. That is why a sudden emotional outburst can scare us and make us stop whatever caused this reaction.

> These are some of the music creation situations where fear can show up:
>
> - While writing lyrics – What if the words are embarrassing?
> - While learning a new tool – What if I never comprehend this as others do?
> - While writing a chord progression – What if I am doing this music theory wrong?
> - While choosing a genre – What if this is not the right beat or instrument for this genre?
> - While publishing a new song – What if it is mediocre, and this is how I will be seen for the rest of my career?
> - While getting new inspiration – What if I cannot finish this track and I will be disappointed?

## How to Stop Fear Getting Between You and Your Goals

Fear can be a frustrating feeling. We acknowledge its existence, and we learn to live with it. Still, sometimes it feels so overwhelmingly difficult to think there could be a life without fear. And due to its hold on us, we can start to believe that it is a source of factual information about what we can do and who we can become.

For example, let us say you might often feel like writing lyrics, but you never dare to take them further than your private notebook. After repeating this action many times, you will only prove that you will stay safe by not doing anything with them. Nothing terrible will occur, but nothing good will happen either. This proof of safety will give your fear more power by enhancing harmful thinking patterns, such as the embarrassment of not being able to do anything with the lyrics or the shame of not chasing your goals.

The reason we get the feeling of leaving the lyrics in the notebook is because of the fight or flight reaction. Because when we experience fear, our survival mode gets activated, and we get the initial response to protect ourselves physically and mentally. Fight or flight mode is divided into two forms: *low-road* and *high-road*.[1]

The *low-road* is an impulsive reaction to fear, where we will react first and ask questions later to get us to safety as soon as possible. This is why beginners in audio might say statements such as "I just do not have a techy

FIGURE 3.11 Illustrated by Emma Holdway

brain". With this, they protect themselves from expectations, separating them from the topic without even trying it first. Or when you get the sudden creative blocks in the middle of a session and get the desire to shut down everything, walk away and forget about the whole thing.

In the event of a threat, the *high-road* means we will consider our options before reacting. We can then first analyse what is happening and whether we have experienced anything similar.[2] For example, a high-road fear response would be seeing a threat, such as a complicated new piece of equipment, that might trigger insecurities in you. Still, you see options to proceed instead of quitting the practise straight away. You might think to take a quick break before trying again or chaining the learning method.

Whether you get the fight or flight reactions often, it is not impossible to find management over it. As much as you can show your mind negative proof, you can also give it positive evidence of your capabilities. The more you are aware of your fear and how it is currently affecting you, the more you can change your low-road responses to high-road. This means you can then choose different approaches to your fear and try ways to get to your goals.

Overcoming numbing fear takes practise, and you need to be patient and give yourself lots of sympathy on the way. The more you actively practise recognising these immediate fight or flight reactions, the more you will get to express yourself. This means you will be able to complete more of your goals and find proof that you are capable of exceeding your expectations.

> Musician and Creativity Empowerment Coach Sara Belle comments:
>
> > When we are caught in this fear there is mainly one thing at play; a concern with the consequences of not living up to perceived externalised expectations of perfection.
> >   We're worried about what people might think, what they might say and what that means about us.
> >   Our fear of failure is trying to protect us from experiencing shame. It has decided that we will certainly fail and therefore we shouldn't try. But what if we decided to shift our focus on thinking about what would happen if we succeed?
> >   To get there, we have to firstly begin to construct our own definitions of success by putting our attention on the process, rather than the product By shifting the desire from immediate perfection to focusing on growth, being able to do a little more today than we could do yesterday, we can dismantle the worry about not getting it 'right' the first time.
> >   We can also dissolve the fear of failure by accepting that trial and error are key ingredients to discovery, innovation and originality in creativity. Our creativity is a journey, how we overcome challenges, based on our unique experiences, is how we get to create work that allows us to show up authentically.
> >   Lastly, we are not our work. Even if we try something out and it doesn't go to plan, it doesn't mean anything about us as a person. It just means that we now have a little more experience under our belts to set us up for another take.
> >   The only failure is to not try – the rest is practise.[3]

## Failure Should Be a Positive Word and Not a Negative One

We have discussed the idea of overcoming the initial fear that might block you from doing things you wish to achieve, but what happens after we try? What if we do not reach the goals we wanted and, instead, experience failure?

Sometimes it might feel easier to not even try to chase our goals as it always comes with the possibility of disappointment. But what if we start to look at it from a different angle. Stop focusing on fixed timelines and goals, and think about failure as an old-fashioned concept that no longer serves us.

I once heard about a girl whose parents did not focus on celebrating grades or other structural success. Instead, they celebrated failure. In her life, taking risks, trying, and being persistent were rewarded. She learned to find success in failure and not the other way around. It is about how you see it, whether you are ready to redefine the word for yourself as a positive and empowering term, instead of something that pushes you down.

Similarly, as with perfectionism, jealousy, and envy, we often put people on a pedestal, picturing their success as painless and without obstacles in contrast to ours. But the people we see succeeding have also felt a failure. For example, Oprah Winfrey was fired from her first job as a news anchor, and Steven Spielberg was rejected from film school three times.[4]

It can be hard to see these stories as comforting because they may feel distant, like these people were luckier than we could ever be. Everyone's starting points in life are different, and the socio-economic environment we were raised in affects how we experience our possibilities in life. Therefore, it can be unfair to compare success stories. Still, whatever might be in your past, it is good to ask whether the definitions you have for success and failure might hold you back from your goals?

As much as we need to redefine the word failure to ourselves, we must also consider what success means. Is success for you a fixed point in the future that you need to get too? Such as a specific grade, an number of followers, or owning a house? Whatever it is, it is good to think about what motivates you to aim for these goals, as success is very personal. What might make you happy and satisfied might not be the same for someone else. Therefore, be careful to define success by what you learned as a child or what others around you value, and aim to find out what makes you feel accomplished.

The question of "Where do you see yourself in five years?" is very common but also problematic. It represents the idea that we should know where to be at a specific time to feel successful. Throughout this thinking model, we put ourselves under pressure to achieve things through fear of failure. But as life is unexpected and we do not have control over the things around us, it is almost impossible to think that the five-year plan will happen as intended. And, saying directly that your goals are possibly not going to happen might sound like negative dream-crushing, but what if it is the opposite? A door to something even better?

We should have big dreams and clear plans, but as mentioned before, maybe in them, we should focus on failure more than success. This means that instead of solely fixating on the end result, we should prioritise the journey itself. By allowing our actions to guide us to where we need to be rather than where we think we should be, we may discover alternative paths to achieving our goals. By expecting potential failures along the way, we can adapt and adjust our plans accordingly, ultimately increasing our chances of success.

This might feel scary, like we are just floating through life without a clear destination. Therefore, instead of using your dreams as a direct map for your next five years, maybe consider using them like a compass. They might point you to one path for a while, but the direction might change. The most important thing is you follow whatever you feel is right, even if it was not in the original plan. With this, your life might initially feel a bit unstable and scary. But, in the long run, it will reduce fear and help you achieve more, while experiencing greater success.

In Chapter 4, we will discuss success and goals in further detail and there will be exercises to practise plan-making and defining your goals.

## Notes

1 James A. Carr. 2015. "I'll take the low road: The evolutionary underpinnings of visually triggered fear". *Frontiers In Neuroscience.* Accessed 15 January 2024. www.frontiersin.org/articles/10.3389/fnins.2015.00414/full.
2 James A. Carr. 2015. "I'll take the low road: The evolutionary underpinnings of visually triggered fear". *Frontiers In Neuroscience.* Accessed 15 January 2024. www.frontiersin.org/articles/10.3389/fnins.2015.00414/full.
3 Sara Belle. (October) 2021. Interviewed by Liina Turtonen.
4 Sebastian Kipman. 2022. "15 highly successful people who failed before succeeding". Accessed 15 January 2024. www.lifehack.org/articles/productivity/15-highly-successful-people-who-failed-their-way-success.html.

## Additional Resources

- Brown, B. 2012. *Daring Greatly: How the Courage to Be Vulnerable Transforms the Way We Live, Love, Parent, and Lead.* Avery.
- Day, E. 2021. *Failosophy.* HarperCollins Publishers.
- I Was Just Thinking. 2023. "Overcoming fear whilst being afraid". YouTube. Accessed 15 January 2024. https://youtu.be/RqKKFwhd5lw?si=1XlBUXKUuZ2CCVi-.
- I Was Just Thinking. 2023. "Navigating failure as a creative". YouTube. Accessed 15 January 2024. https://youtu.be/rvOidsKCYKk?si=v-Q2gdaqTc1hwJ0U.
- Manson, M. 2016. *The Subtle Art of Not Giving a F\*ck: A Counterintuitive Approach to Living a Good Life.* HarperOne.
- Robbins, M. 2020. "If you're afraid to fail, WATCH THIS". YouTube. Accessed 15 January 2024. https://youtu.be/e9pTjkXviT8?si=RAkNW8Bu5MZba-d7.
- Shetty, J. 2020. *Think Like a Monk.* Simon & Schuster.

# Workflow Theory 4

## Keywords

- plan
- workflow

## Plan Your Creativity

In this book, we have worked through the reasons for feeling insecure about our creativity or why the creative processes so often get blocked by our minds. This chapter will focus on plans and workflows, what they mean, and how they can help us overcome our creative struggles.

As a neurodivergent person, it has always been essential to dissect any process to its core, to understand it fully and to be able to utilise it. When at school, I remember asking the teachers so often, "but why?". Whether it was maths, biology, or a music lesson, I wanted to know why I was doing the things we were taught, so that it would make sense to me. But to my disappointment, the answer was usually, "It is the way it is. Just memorise it. The whole reason would be too big for you to comprehend".

As an adult, I have realised that answering *why* does not need to be complicated or even highly detailed. It is just a summary of all the reasons we should be motivated to use the information we have gained to take us forwards. The same goes for our creativity. Without fully understanding why we do the things we do and dream of particular things, how can we ever know what move is right for us to take next?

Therefore, I believe our creative workflow does not only consist of the practical steps of which instrument you first use in the composition process, but from much further back. It starts with you asking yourself, "But why?".

Understanding why you create art and where you wish to go with it can lead you to your core values, revealing your definition of success. After this, we can utilise plans and workflows to assist our creative organisation. They are essential in breaking down dreams, visions, and goals into smaller, more tangible pieces. Having a guide to direct us from A to B makes us feel more secure in our journey, reducing anxiety and fear about doing something outside of our comfort zone. Similarly, following the steps you have created for yourself helps us stay on top of our insecurities while chasing dreams that can otherwise seem too big and overwhelming to achieve.

- *Plan* is a structured dissection of your aim guided by your core values. It is a manuscript that lays out every part of your goal into a clear and realistic path. Plans allow you to see your goals as a big picture, like a map.
- *Workflow* is the practical steps towards your goals. In this book, we will use the word *workflow* to describe a detailed guide that gives step-by-step instructions on getting from A to B. It is a pre-planned guideline used as a way to navigate a project.

In the next chapters, there are examples of both plans and workflows, and you will learn techniques to overcome your insecurities by using both for more balanced creative confidence. They will help us organise our complicated and sometimes messy brains in a very substantial way. Many of us dislike rules, but a structure gives us comfort. Having an idea of what we can expect from the future can reduce our stress and fear, helping us focus more on the journey instead of the outcome.

## Plans: Organise Your Goals and Deal With Expectations

One of the hardest things in creative practice is to be honest with ourselves on what we truthfully expect from the outcome of our art. This is because finding out what brings you happiness and a feeling of accomplishment is often affected by the circumstances and the *space* around us (see The Quartet of Creativity in Chapter 1). Therefore, we can get blinded by the values and goals we grew up with, making it harder to see what we truly desire for our outcomes and expectations.

Unclear expectations and goals can blur the *why* in *why we create art* but also set us up for feelings of failure when we create. Therefore, plans and

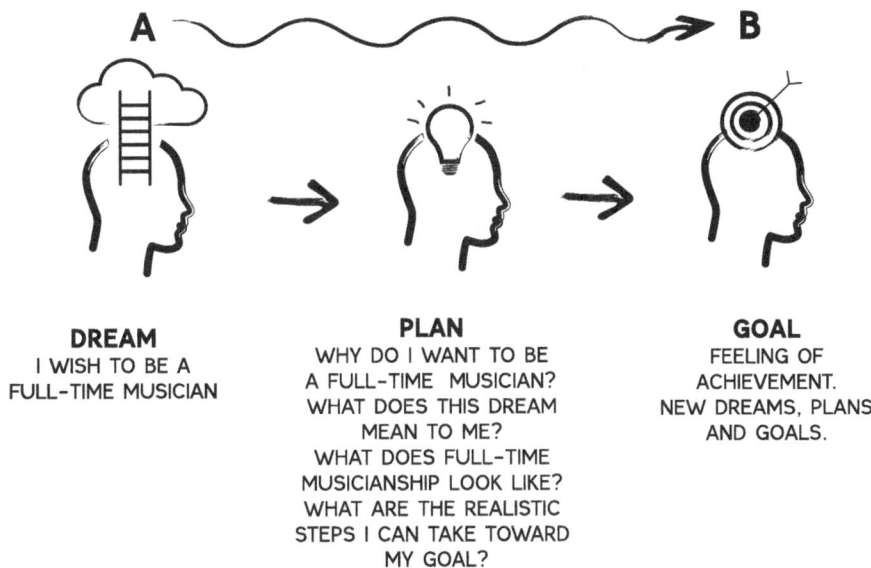

FIGURE 4.1 Illustrated by Emma Holdway

workflows can help us to make our goals clear and help us make the journey into these dreams a reality.

For example, you might wish to be a successful musician. In this case, consider the following:

- What is the measurement of this success? What does being a musician means to you?
- What is your starting point? What are your skill level and financial situation?
- How much time do you have to dedicate to this project?
- Is the musician lifestyle for you? Would you be okay with being famous?
- Will you be willing to work for this dream, be persistent and patient?
- Why do you wish to be a musician? What motivates you for your career? Is it money or validation from your family or peers?
- Who do you want to impress the most, others or yourself?

There are no right or wrong answers to any of those questions. Whatever dream you wish to chase, all of these questions are valid. A dream is never too big to have, and by being honest and realistic about your journey towards your goals, the better your chances are of finding the right path. With a better understanding of what you wish to achieve, you might discover new dreams. It is important to give yourself the freedom to explore these new aspirations without feeling like a failure if you change your mind. In fact, pursuing these new goals could lead you to something even more fulfilling than your original dream.

I have had several big dreams throughout my life, and I have changed them three or four times. Initially, I thought that changing my dreams meant that I was giving up. However, I now view it as an opportunity for personal growth and expansion. I never lost anything by changing my dreams; instead, I have only gained. My first big dream was to become an actress, which I held onto for 21 years. I felt defensive whenever someone suggested a different career path for me because this dream had become a part of my identity. Later on, I discovered my love for audio and music production, and my main goal was to become a well-known audio YouTuber. But now, during the years of writing this book, I have realised that my future path might look different again as I have grown and changed. Change can be scary, but I have learned to trust my gut feeling and believe that if I work towards the things that make me happy, they will happen eventually. All I need is patience, and the motivation to continue will come from the trust that things will happen if I keep trying.

As mentioned before, plans are just a compass guiding us in the right direction. Whatever you wish to be, a pop star, sound designer, DJ, YouTuber, artist, and so on. Start by taking a step in one direction and be open to wherever it will lead you. From time to time, check back to your plan, refresh it, or even change it entirely, but be honest with what you truly want.

## Workflows: Finding Flow and Putting Our Creativity Above the Technology

As discussed in Chapters 1 and 2, creativity in music production is affected by several factors in surrounding industries; technology, the tools we use and how we interact with them, music industry pressures, and the mental game of our creative processes. Navigating all of this is not easy while chasing the flow-state and a total enjoyment at the same time. We all know how big tasks feel easier to approach when they are separated into smaller steps. This allows us to focus on one thing at a time without stressing about the whole journey at once. Workflows are the same thing, a structured guide to help us practically and emotionally achieve our goals, one step at a time.

Similar to plans, in creative workflows, it matters how you approach your outcome and what expectations you will have for it. Let us say you wish to make a song that does well in the pop market. If this is the starting point for the track, it should also be reflected in the workflow. This means you would need to have a realistic plan for making it. For example, whether you have the skill set to create a track that will get the attention required, or will you need to plan time to practise these skills first. Or perhaps you do not have much visibility online or in the music industry yet. In this case, it might be worth giving attention to growing that before creating high expectations for the track's reach. But managing your expectations makes your dreams realistic, tangible, and even more possible. When you know what to do to achieve it, it is only about creating a practical workflow that puts that plan into action.

As music producers and audio creatives, we can find ourselves at the crossroads of technology, audio, music, and creative industries. Using a music studio as your primary creative tool differs from using, for example, a single pen. The workflow is clear and straightforward when an artist is given a pen to create a picture with. Whereas offering a musician a computer, the shift in artistic workflow blurs and becomes more about what the tool provides us, rather than what we wish to accomplish with it. For example, let us picture a very regular size home music studio. It has a computer with music-making software, some VSTs, and additional apps; alongside there might be a couple of controllers, hardware, and analogue instruments. Now consider how many creative approaches and options this amount of equipment holds inside of them – endless.

This is where our creative confusion and blocks often start. Our creativity can easily be short-lived due to the sheer number of options for both approach and sound. Therefore, insecurities can arise when our fragile creativity is put against infinity, taking away the focus from why we started to create in the first place. For example, concentrating on the tools, techniques, and our lack of knowledge more than the goal we had in mind. At least with a pen, it is clear how it will assist us as a tool. With a computer, it is never that clear.

Therefore, we can use creative workflows to help us overcome these overwhelming feelings with both the fear of blank canvas and technology. So whether you are a beginner who might find insecurities about knowing where to start or an advanced user looking for more structured inspiration, workflows can help you to focus and enjoy your process, without the pressures of the outcome. Having guidelines and limitations will give a starting point: a puzzle to be solved.

Considering creative work as a puzzle is an exciting and inspiring thought. Limitations and structure are the core of a workflow, but to solve what? If we think about our outcome as a problem, we might present ourselves with a negative starting point. Maybe even with a self-destructive mental space which can encourage perfectionism and feelings like envy, frustration, anxiety, and fear. Milton Mermikides, a composer and producer, explained how treating our creative projects as a puzzle can be beneficial in an interview:

> One thing that has helped a lot is the idea of a music project as a sort of puzzle instead of a problem. Puzzles are always fun and have a solution because they are about play. But for example, if you write a piece on a silly idea such as "four big and three small notes". And then that is the puzzle to interpret how you like. Suddenly you have ideas already, and you start to think, what does big mean? What does small mean? How many times can I repeat them? So that is a little like a vignette. And that is how the puzzle sort of becomes the creation, like a pencil drawing itself.[1]

When you find direction for your creativity, you have more chances to feel inspired, as inspiration comes from the excitement of solving a vision you had. For example, imagine that you have created a 10-step workflow to make a house song. The goal and expectation are to finish this track. You will start by adding drums, then bass and chords. Thus far, you have purely followed the instructions the workflow has given you. But suddenly, the chords have shown you a surprising melody pattern which makes you feel something. This melody intrigues you, and you wish to see where it leads you. It was not in the plan, but you found inspiration through it. Or maybe the inspiration dies down at one point, and instead of feeling disheartened about it, you refer back to the workflow and use it as a guide for the next creative decisions.

In Chapter 1 and 2 we discussed creativity and achieving flow-state, addressing how it is a combination of many factors, varying from confidence to interface design and user experience. To find our way into the flow-state, we need to consider how we interact with our creative tools, alongside our psychological thinking models, such as *divergent* and *convergent thinking*. And, as our lives are busy, meaning we often do not give our creativity the time it deserves, we need efficient time management to ensure we have the head space to reach for new ideas and process them while still learning new techniques. But as mentioned in Chapter 1, it is recommended that you do these at separate times, using *time blocking methods*. Examples of workflows on how to do this practically are shown in the next chapter.

Therefore, you can find comfort and a feeling of security in the structure by understanding and organising your creative process from the tools you use to time management, learning and innovative thinking models. A chance to empty your mind, trust the process, have confidence in yourself

# Workflow Theory

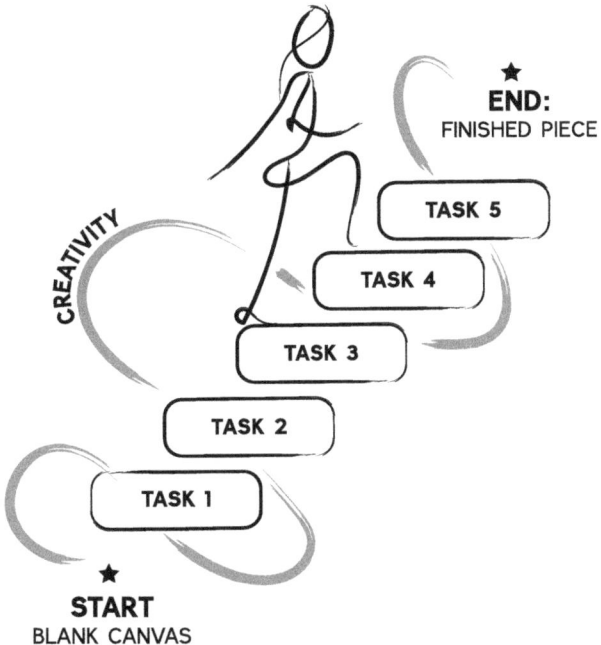

FIGURE 4.2 Illustrated by Emma Holdway

and your skills, and trust that there is no failure in creativity. From this enjoyment and comfort, you can find inspiration to solve the puzzles of your imagination, leading you to the *flow-state*. Flow means feeling a moment of clarity in our vision, where our fears and insecurities do not filter the ideas coming to us and feeling at ease with every thought that comes through us. In the next chapter, we will look at techniques to create workflows and plans, and I will share samples of ones that have proven to be helpful for my students and myself.

## Note

1 Milton Mermikides. (August) 2022. Interviewed by Liina Turtonen.

## Additional Resources

- Allen, D. 2015. *Getting Things Done: The Art of Stress-Free Productivity*. Penguin Books.

- Csikszentmihalyi, M. 2004. "Flow, the secret to happiness". TED Talk. https://www.ted.com/talks/mihaly_csikszentmihalyi_flow_the_secret_to_happiness.
- Kelley, T. and D. Kelley. 2013. *Creative Confidence: Unleashing the Creative Potential Within Us All*. Currency.
- Tharp, T. 2006. *The Creative Habit: Learn It and Use It for Life*. Simon & Schuster.

# Get to Your Goals with Workflows 4.1

Chapters 1 to 3 of this book aim to give awareness on why we might feel insecure and blocked with our creativity. In the previous chapter, we discussed workflows and plans and how they can help you with your creativity, fulfilling your creative potential and reaching your goals. In this chapter, you will learn how to create plans and workflows of your own. The techniques are aimed at music producers, but they can be applied to the artistic workflows of any individual.

These methods come from years of analysing my creative workflows and the observations I have made from my students' creative processes as an educator. That is why they are completely unique for this book. After applying these methods in creative practice for years, I have also adapted them to my everyday life—something I highly recommend trying out. To make the most of them, the artist must be willing to open up to the challenge, which might sometimes take a hint of vulnerability and trust in the process.

Based on the workflow theory explained in the previous chapter, I have divided the workflows into three sections: *mind workflow*, *life workflow*, and *action workflow*. We can picture these three workflows as a pyramid, and in this chapter we will explain each section individually.

## The Pyramid of Workflows

At the bottom of the pyramid, we have *Mind Workflow*, the basis for our creative confidence (refer to Figure 4.3). If we have an awareness of what our main dreams and values are, as well as the leading causes of our insecurities, we can find balance and motivation in receiving our full creative potential.

## 136  Creative Confidence and Music Production

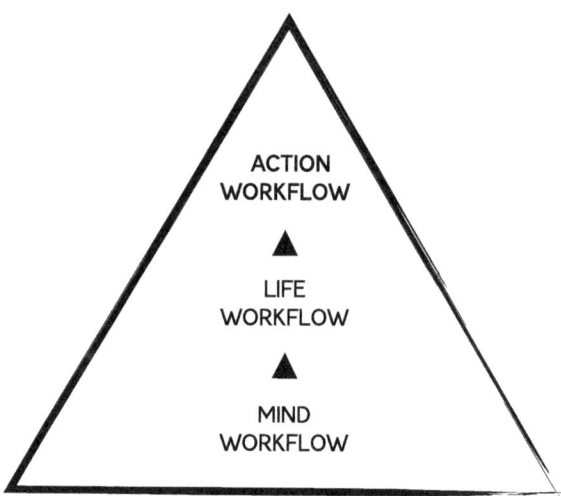

FIGURE 4.3 Illustrated by Emma Holdway

This is where you can use both plans and more detailed workflows to understand your goals, definition of success, and what drives you towards these dreams. Focusing on this section before moving to the other two is recommended, as it will make further planning much more straightforward and clearer.

The following workflow is *Life Workflow*. It is the more detailed action plan to make all the goals accessible and tangible from the *Mind Workflow*. You can use these techniques to organise your creativity around your busy life, ensuring no goals are missed and finding the most efficient ways to do your projects while maintaining your creative confidence.

At the top of the pyramid, you will find the *Action Workflow*. These are the type of detailed structures we associate workflows most with. This is where you will guide yourself through the practical process of creating and deciding what techniques and tools to use while setting the framework for the project.

Next, we will have a detailed look into all of these workflows, and you will be given instructions on how to implement them in your own life.

## Mind Workflow

Here are two suggested Mind workflows you can create for yourself: *Big Picture* and *Self-dialogue Workflow*. These are both for you to design according to your needs, helping you to clarify the journey to your goals from A to B.

The *Big Picture Workflow* is a planning process that helps you to clarify your values, expectations, dreams, and goals. As discussed in the previous chapter, managing your expectations is crucial to achieving your goals more quickly and taking care of your confidence along the way. This is where you will define for yourself what those expectations are.

*Self-dialogue Workflow* is a little bit more unusual. Its structure is not based on a plan but more on a strategy for the moments you might feel lost in your thoughts and insecurities. It is formatted as dialogue and a practical tool to help you overcome creative blocks. It will be there as a reminder to guide you through some of the significant insecurities you might battle with.

## *Big Picture Workflow: Planning for Goals, Dreams, and Success*

The planning for your creativity should start from the beginning. This method will help you figure out your ultimate goals and structure your dreams. The more measurements you can apply to your goals, the more tangible and possible they become. This workflow can seem like a lot to plan at first, and it might take a while for you to do, but I recommend not skipping this phase. As shown in Figure 4.2, we need to define the bottom of the pyramid first before building on top of it.

In this workflow you should focus on defining the core values you have for your life that motivate your goals and aspirations. These are some of the main aspects that make who you as a person, what you believe in, what you think is important in your life and how you wish to spend your time. Try to be honest with yourself when doing these exercises, and leave others' opinions and influence outside of it. Remember, you can always change your mind. That is why it is good to repeat these workflows time after time, adapting them to your growing and expanding mind.

**Here Is How You Create Your Big Picture Workflow**
(Refer to Figure 4.4.)

This workflow is great for when you are lost in your direction and need little hope in your future. Doing this helps you push away all the things in your life you have no control to affect or change and enables you to focus on all the things you have power over. The example below is meant to help anyone navigating the music industry, but this is an excellent tool for you to use in any area of your life.

FIGURE 4.4

Illustrated by Emma Holdway

You can do this exercise on a regular empty document, but I prefer creating a Sheet or Excel document to make it more straightforward.

1. Name all the goals and dreams you might have in your life. Remember to be honest, and do not think there are any dreams too big. Remember to add some of your big dreams that might not be involved in your artistic goals. It might look like this:
   - Publish a solo album
   - Learn to mix and master
   - Be self-employed and be a full-time musician
2. Break each of these dreams into smaller sections, defining what the plan looks like on a practical level. Think about the time scale and the proportion of this dream. This might look something like this:

- Publish a solo album:
  - In the next 1–2 years, around seven songs, composed, produced, mixed and mastered fully by me.
- Learn to mix and master:
  - In the next six months using Ableton Live.
3. Now let us see practically what we can do for each goal. Go through each of the subcategories you just wrote down and think about what it takes for you to get to this objective. It could look like this:
   - Publish a solo album:
     - In the next 1–2 years, around seven songs, composed, produced, mixed, and mastered fully by me.
       - I need to give myself time to compose and write these songs. Attend or arrange a songwriting camp to get inspired by working with others.
       - Use step-by-step workflows and time limitations to complete a new song every week.
   - Learn to mix and master:
     - In the next six months using Ableton Live.
       - Find an online course: budget $120.
4. Keep on dissecting the plans in even smaller and practical steps.
5. Some of the following steps might already seem very clear at this point, but the dreams are not tangible enough if you have not added anything to your calendar yet. Therefore, the next step is creating a separate action plan for the next weeks, months, and years. For this stage, you can start another empty document. Start by creating a timeline of the year, adding months and weeks. Drop the tasks you defined in step 4 of this exercise into the time grid. This might look like the example shown in Figure 4.5.

As mentioned above, completing all of these steps might take you a while, but it is well worth dedicating some time to. Once you have the main structure down, you can keep returning, modifying, and adapting it to your ever-changing life.

## Self-dialogue Workflow

The simplest way to create a *Mind Workflow* for yourself is by writing down the fundamental insecurities you might face while doing your creative practices and giving yourself an already designed answer that you can refer back to whenever you need to. It is a written script with the little voice in your head that can be harmful to you but is only based on imagination.

Imagine a typical scene where you might need to use this workflow:

You are working on a song in a music studio when suddenly, out of nowhere, you get the strong reaction to turn off all your equipment and do something else. You recognise this as a fight or flight reaction to something that came to your mind while making music. You are unsure what it was but feel distracted from your inspiration. This is the moment you might refer to your Self-dialogue Workflow, which might be printed out as a PDF document, written down in your notebook, or structured in a cloud-based document.

**This Is What Self-Dialogue Workflow Might Look Like**
**The little voice in my head:** This song will never be good enough. Even if you release it, it will be laughed at, and people will see how bad a producer you are.
**Me:** My goal is to finish this song. Whether it be the hit song of the year or a little tune, I will decide whether to publish it later. No matter what I create, there is no failure in it. I make music because it makes me feel good and have fun. No one knows all the techniques, and I am proud of how far I have come in my audio learning journey. I am not here to impress others, only myself.
Add as many of these discussions as you like.
This might feel funny at first to write down, but having it there to remind you of what is real and which fears you have imagined will help you soothe the voice that creates self-doubt. Similarly, by assembling this *Mind Workflow*, you will find a way to comfort yourself first-hand, meaning you do not need to look for external validation immediately, but continue your creative practice.

## Life Workflow

The second workflow in the pyramid is *Life Workflow*, a practical way to organise your creativity in your busy life. After you know what your ultimate goals are, as defined in the Mind Workflow, this stage gives you the tools to organise your life and your creativity so that they will support each other. This is based on *time-blocking* practices and psychological thinking models: *divergent and converged thinking*. In addition to this workflow, learning new skills is given its location in your schedule. To read more about how these concepts can affect your creativity, refer to Chapter 1. Here are shorter reminders of each keyword and its meaning:

- **Time-blocking:** A technique that helps you block time from your schedule for all you need to accomplish during a day, week, or month. Similarly, the way employees or school would give you a plan on where

and what you need to do at a specific time, you can create a schedule for yourself. Planning your free time and creative activities will make your time management tasks more approachable.
- **Divergent thinking:** Reaching out for ideas and letting them go through you without judgement or further analysis.
- **Convergent thinking:** The processing of the created material and ideas. This is where you analyse and apply what you have thought in divergent thinking into a creative structure.

Having a clear idea of when and what you will create will reduce the stress of your everyday creative time management obstacles while giving you a guideline for more efficient practice that minimises your creative insecurities. That is why this workflow can assist you also if you struggle to find time for creativity or overcoming your creative blocks.

By applying this method to my personal and work life, I have found a healthier relationship and balance in all that I do. I have found it easier to give time for my music, found healthier balance between procrastination and perfectionism, and reaching my goals faster. This has also made me enjoy my creative time more, allowing me to entirely focus on one thing at a time, making it harder to think about other responsibilities while creating. With this workflow, you can give your creativity the time it deserves.

## *Divide Your Time into Three: Divergent and Convergent Thinking and Learning*

Here is an example designed for anyone with a full-time job and family or other significant commitments. It demonstrates how even short creative times, such as one hour a day, can already seem more tangible and easy to approach:
Assets for the workflow plan:

- **1 hour of divergent thinking.** This can be playing with a new instrument, finding a collection of random samples, or going for a walk and writing down whatever comes to mind (journaling). Do whatever inspires you at that moment. Remember not to judge any of the ideas that come to you.
- **1 hour of convergent thinking.** Look through the ideas, samples, and music you have created in your divergent thinking session. Start fitting them together and see if you can find structures and patterns that please you.
- **1 hour of learning.** This is the time for you to research your tools and expand your knowledge. By having a separate time for it, you do not get overwhelmed by new things while creating.

Your weekly workflow plan could look something like what is shown in Figure 4.5.

## MY WEEKLY CREATIVITY PLAN

| MONDAY | TUESDAY | WEDNESDAY | THURSDAY | FRIDAY | SATURDAY | SUNDAY |
|---|---|---|---|---|---|---|
|  |  |  |  |  | 9 am<br>1 hour of learning |  |
|  |  | 5 pm<br>1 hour of Convergent Thinking |  |  |  |  |
| 8 pm<br>1 hour of Divergent Thinking |  |  |  |  |  |  |

FIGURE 4.5 Illustrated by Emma Holdway

Notice that you can use this workflow in any way you find suitable for your needs and goals. For example, as a full-time freelance musician, producer, content creator, and author, I use this workflow in many different ways.

**Here is an example of how I arrange my time:**
Monthly project time-blocking: As a content creator, I must constantly develop new ideas while finding time for my music, practice, and other free time. This is why I often split my year into three or four blocks, allowing me to focus on different projects for a couple of months. I can write my book for two months at a time, then create videos and social media for another couple of months, doing the same with my music.

During these months, I divide learning, divergent, and convergent thinking into their days. For example, this might mean I use one day to study what topics are trending and one day to come up with ideas for my YouTube channel and what to film. This gives me a guideline and structure to follow, giving practical steps to finish each task.

As a self-employed person, it is sometimes hard to follow the planned schedule, but this approach helps me manage stress. I trust that each task will be done on time, as everything has its dedicated time slots.

Get to Your Goals with Workflows    143

## Action Workflows

Action workflows are the guidelines for when you do not know how to start, when you get blocked, and when you need to finish your song. They are clear, step-based workflows that break down any task into simple actions. Here I show three *Action Workflows* you can create: step-by-step, puzzle-solving, and finishing routine workflows. In the next chapter, I will share with you five more step-by-step workflows that can help you with your creative practices and give you a starting point when applying workflows to your creative routine.

### *Step-by-step Workflow*

This is the most commonly known workflow, a plan for a personal creative process structure. It is where you design which instruments and tools you will use and in what order. Often this can come to you naturally, but giving our workflow space to form naturally can sometimes cause more creative blocks than inspiration. This is why giving yourself a creative action workflow can be extremely helpful. Read more about the theory behind workflows and plans from the previous chapter.

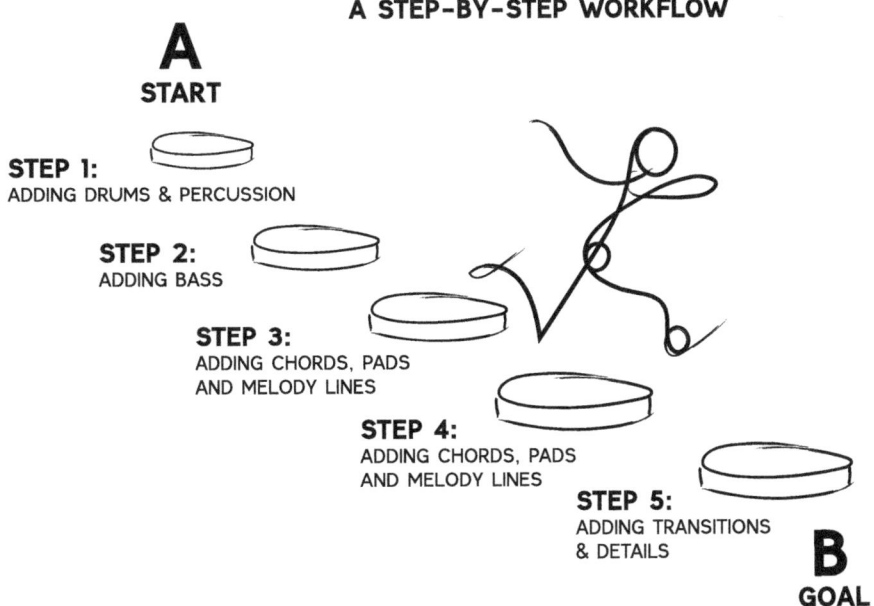

FIGURE 4.6  Illustrated by Emma Holdway

Step-by-step workflow plans can be designed from several different points of view. Here are some of the ways I usually approach forming these workflows, both for myself and my students:

- **Genre-specific.** Whether you are a beginner or a professional music producer, having guidelines for a genre can be extremely useful. For someone just starting, it can be beneficial to get directions on what bpm (beats per minute) they should begin with, what drum pattern to apply, and what type of instruments they should use. Whereas for someone knowing the genre, these guidelines can give inspiration and a new perspective.
- **Limitation focus.** Applying limitations in your step-by-step workflow makes it easier to focus on the creative process more than on the outcome. For example, by limiting your process only to two or three tools, you are creating yourself a puzzle to solve that increases the exploration, element of surprise, and interactivity with the chosen tools. Or, by giving yourself a time limitation, you can overcome the fear of failure by changing the expectations from "creating something amazing" to "creating something in the time frame I was given". In this scenario, it is much easier to let yourself create almost anything, as the responsibility of the expectations are defined by the challenge and not by you.
- **Music theory-oriented.** This workflow is great for anyone who wants to learn more about music theory or finds it useful in their creative process. In this workflow, the song is structured based on harmonic, rhythmic, and arrangement techniques. This is also great for any musician, regardless of what equipment or instrument they might be working on.
- **Structural Inspiration.** This workflow is based on building a song or track using additive and subtractive methods. The idea is to find building blocks from 3–4 elements first, make these to an arrangement that works well together, and then add necessary details. This approach is beneficial for anyone working in a DAW.
- **Collaboration perspective.** Collaboration is one of the best ways for us to explore new perspectives and find inspiration. There are millions of ways you could collaborate with another musician, but in the workflow example in Chapter 4.2, I will show a suggested approach, which can be done in real life or remotely.

All of the above workflow approaches were invented and tested by me, but they can be applied or changed depending on what works for you. Therefore, I challenge you to think of other ways to approach workflows and create templates you can share with others. If you wish to share your ideas with me and the community, please check out the link below. Examples of each of these step-by-step workflows can be found in the next chapter.

## Puzzle-solving Workflow

Creative processes are often about finding solutions to puzzles we create for ourselves. This is why sometimes, in the midst of our workflows, we might encounter problems that make us stuck. These might be technical issues, such as the sound does not come through, or maybe your computer is overloaded and gets frozen. These issues are common for anyone working in audio, and whatever stage you are at in your career, these issues do not get any less annoying. That is why it can be good to prepare for these situations and prevent them from stopping your creative flow.

These tips will help you, especially if you are a DAW-based music producer, but also if you work in studios or with analogue gear:

1. **Make a Troubleshooting List**
   Create a physical list of all the possible issues you might face with audio routing. Having a clear document of the signal flow in the studio can help, especially if the setup is shared with someone else or if you are a beginner needing a reminder on why sound might not be coming through. It is also practical to write down other similar possible points that might confuse you in the middle of a session. With experience, it is customary to create one of these lists in your mind, but sometimes writing these possible obstacles down into a reference document can save time when needed.

2. **Try Not to Get into the YouTube Tutorial Trap**
   We all know the feeling when we are in the mid of the session, and something we wish to do does not work, or we need to understand some features of a tool to achieve our goal. But to get this information, we go to YouTube and look for answers. As much as this can be helpful, it can also trap us into the depths of the internet. This distraction can ultimately make us lose inspiration and make us feel overwhelmed or insecure as we might start to compare our work to others. As mentioned in the *Life Workflow* section of this chapter, try to separate your learning time from the creative moments. If you do not know something at that moment, maybe it is okay, and you can concentrate on the techniques you are familiar with instead of gaining more knowledge.

3. **Save a Copy**
   It is common to practice keeping all your files saved and having backups on everything, but having copies of your sessions can also help with creative blocks. When you feel like you have no idea how to continue with your song and are confused with its direction, save a copy of it. This means you can look at the duplicate with fresh ears and eyes and not feel too precious

or attached to it. Start removing elements, adding, cutting, looping and have fun turning what you had into something completely unexpected. So, you can always continue with the duplicated file or take elements of it and return to the original track.

## Finishing Routine Workflow

Knowing when our tracks or songs are ready can sometimes be challenging. Especially when we have listened to them for hours, deadlines are closer, or we start to feel tired with the process. This is when we can have a workflow for the last push.

You can create your routine, but this is a workflow I like using when I need to get work finished efficiently:

1. Listen to the track, eyes closed, without doing anything simultaneously. Make sure not to change anything on the project while listening. Just focus on the sound.
2. Write down a list of things you still wish to change on it.
3. Do the changes.
4. Have a break. This can be a coffee in another room, a walk or even a week between sessions.
5. Repeat the same sequence. Listen, write notes, make the changes, and have a break.
6. Repeat as many times as needed until you have no items on the list after listening to it. This is when you know it is ready.

## Additional Resources

- Find more about creative workflow and where to share your own tips here: https://www.creative-confidence.com/.
- LNA Does Audio Stuff. 2020. "40 ways to finish tracks in 12 min – Music Production Tips". YouTube. https://youtu.be/iSYOEeihKcc.
- LNA Does Audio Stuff. 2022. "How to make house track in 10 steps". YouTube. https://youtu.be/EDiZrTccfyg.
- LNA Does Audio Stuff. 2022. "How to make techno track in 10 steps". YouTube. https://youtu.be/8om2uak_FKU.
- LNA Does Audio Stuff. 2022. "5 ways to overcome creative blocks in music production". YouTube. https://youtu.be/1LxF7OzkOuA.
- LNA Does Audio Stuff. Download workflow templates from. www.lnamusic.com/shop.
- Terhinator, DJ. 2022. "Ready to finish more music? 3 top tips for better time management". Blog Post.

# Workflow Examples 4.2

## Keywords

- music theory
- musical genre

In this chapter, I will introduce five examples of *Action Workflows*, which were explained more in detail in the previous chapter. These workflows help give a starting point and a guide for your creative process. Remember, these are not rules but a compass to your inspiration and finding your flow. If you get stuck, refer to the workflow and follow it until you feel inspired again.

All the workflows are designed to work for anyone with any DAW (Digital Audio Workstation), musical instruments or analogue gear. These workflows are recommended to keep as an example, so you can start to design your workflows. Try different approaches, or even specify them to fit your unique equipment. The main point is to find ways to move on from creative blocks while making space for your creativity and enjoyment.

## Workflow 1: Genre-specific Approach

Let us use house music as an example. This workflow can be designed to follow any genre. Simply dissect its characteristics and put them into a straightforward guide like this.

**House track in 10 steps:**

1. **Step 1 – Vocals:** Soulful vocals usually work very well with house. Find vocal samples or create two to three vocal lines that will determine the

song's scale. At this point, just add them to the session, and we will arrange them later.
2. **Step 2 – Hi-Hats:** If in doubt, use 909 drum kit sounds. You can also choose a couple of different hi-hat sounds and create a pattern using all of them to add variation. Adjust the velocity and expression controls to add more "human sound" when using electronic instruments.
3. **Step 3 – Core Drum Beat:** You can use 909 drum kit sounds if you need a starting point. Start with a four-on-the-floor kick pattern, then add snare and clap every second and third beat of the bar.
4. **Step 4 – Synth/s:** Find or create a nice synth pluck-style instrument. Create a 4-bar loop with a chord/notes complementing the scale of the vocals and add an arpeggiator to create a rhythmic pattern.
5. **Step 5 – Chord Progression:** Upright piano sound always works with house tracks, or try a soft keyboard sound. Find chords that suit the synth and vocals, and apply them in a 3+3+2 pattern. This means the first chord will take the three first 16th notes of a bar, the second chord the following three 16th notes and the last chord the last two 16th notes.
6. **Step 6 – Bass Line:** For this, you can use a simple single oscillator bass sound with any wave shape you find inspiring. Copy the 3+3+2 pattern you applied to the chord progression, apply bass notes to complement the scale and apply them an octave lower than the chord progression. Add some swing to the notes to improve the groove. Now adjust the tempo somewhere between 110 to 140 bpm, where it sounds suitable to the rhythm you have created.
7. **Step 7 – Arrangement:** An example arrangement for a house song would be: ABCB. This means verse, chorus, bridge, and chorus. You can also add an intro and outro if you feel they are appropriate.
8. **Step 8 – Effects:** Apply your favourite effects to the vocals and instruments. Delay works amazingly well on house vocals and pluck-instruments. You can also use filters to create variation and tension in the arrangement.
9. **Step 9 – Swells and Lifts:** For example, automated arpeggiator synth lift, white noise with LFO, synth with LFO/Side chain and an Auto Filter. You can also create a simple swell for lifts and drops by reversing a cymbal and adding effects to it.
10. **Step 10 – Fills and Ear Candy:** Add quietness before and/or after choruses and verses. These quiet moments are nice to fill with drum fills, vocal chops with effects, stretched and filtered audio, or any other creative and fun-sounding effects.

To watch this workflow in action, watch this video where I make a House track in 10 steps: https://youtu.be/EDiZrTccfyg (LNA Does Audio Stuff, YouTube, 2022)

## Workflow 2: Limitation Focus

This workflow focuses on applying limitations to enhance your exploration, an element of surprise, and limiting feeling overwhelmed by the options.

**Time-limitation workflow example:**

1. **Choose a realistic goal for yourself:** This could be, for example, creating a short four-track production with chords, drums, bass, and vocals. The aim is not to make a full or ready song but to create a sketch of an idea.
2. **Plan your tools:** Think about what equipment you wish to use in this challenge. Consider using tools you are very familiar with and feel comfortable using. Try to limit the options by having only a couple of synths and a small selection of drum kits ready to hand, so you will not use the time to choose sounds.
3. **Eliminate distractions:** Make sure to be in a quiet space where you feel like no one can disturb you. Also, turn off all notifications and put your phone away.
4. **Time challenge:** Put a timer on for any time that feels comfortable for you and what you have time for. This could be 15 min or 4 hours, more or less.
5. **Overcoming blocks:** While doing the challenge, remember to be open to change, new ideas, and different directions. When working on a song or a track, it is important to let it evolve naturally without trying to control the outcome. Trust your intuition and follow the golden rule of improvisation: say yes to each idea that comes to mind and build on it with an "and…" Then, think about what you want to do next. There are no bad ideas; trust the process, using your puzzle-solving skills to figure out which direction the track should go.
6. **Repeat:** Schedule this same exercise on your calendar as a weekly thing you do. Save all the sketches you create so that you can return to these ideas at a later date.

**Other limitation workflow ideas:**

- Create a song using only tools you are not familiar with.
- Use only one instrument.
- Create a whole song using one sample.
- Use a tool you hate.
- Let someone else decide your goal, your tools, or the topic.
- Work in a genre that is unfamiliar to you.
- Start with an instrument you do not know how to play.

- Create five ideas in three minutes: put a timer on and start creating. When three minutes is over, start a new project and do the same again.
- Play with your eyes closed.

## Workflow 3: Music Theory-oriented

This workflow is based on music theory and is helpful for anyone seeking inspiration from harmonic and rhythmic structures rather than a sound-led approach.

**Music theory workflow example:**

1. Choose a scale you wish to work in: For example, A Minor.
2. Create three melody lines using only three to four notes from the scale.
3. Pick a chord progression in A Minor. To challenge yourself, use the roman numerals system and pick a progression at random. If you are not familiar with the roman numeral system, try using a chord progression generator online or pick one of the following progressions: i/III/i/i – Am/C/Am/Am, VI/i/iv/i - F/Am/Dm/Am, VI/i/VI/v - F/Am/F/Em, i/v/VI/VI - Am/Em/F/F.
4. Next, apply the bass line. Work on the pentatonic scale, and in this instance, that would be Am Pentatonic. Create a bass line running up and down in the scale, finding a suitable pattern that matches the harmonic structure of your chords.
5. For the drums, choose to work with either linear or polyphonic drum patterns. Linear means that none of the drum sounds play simultaneously, whereas in polyphonic patterns the sounds can be played in layers.

## Workflow 4: Structural Inspiration

In this workflow, we focus on building a song using simple structure methods and additive and subtractive arrangement tools. The idea behind this creative process is the same as building a house; we must first lay a strong foundation, make the frames, and finally decorate it in our unique style.

**Building a House – Workflow**
**Foundation:**

1. Use a short time to create a collection of samples, sounds, loops, and clips of everything that inspires you. For example, this can be chords, lyrics, melodies, bass lines, beats, foley, and soundscapes.

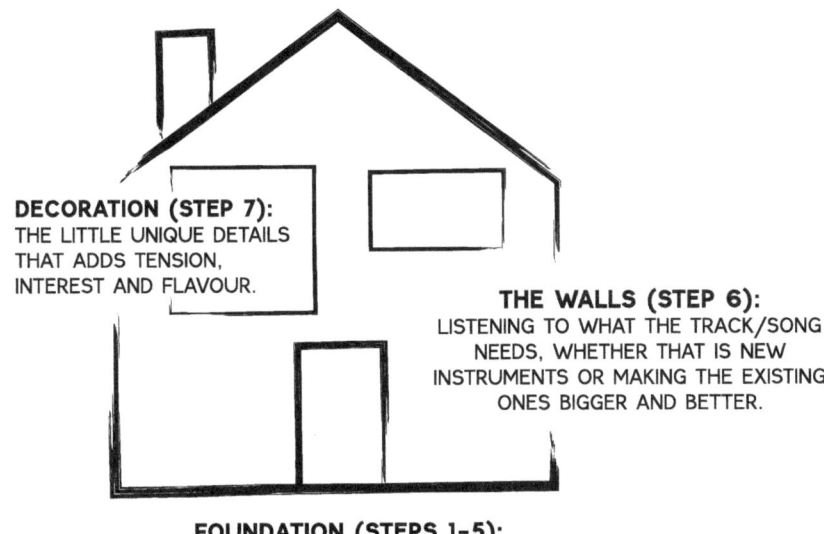

FIGURE 4.7 Illustrated by Emma Holdway

2. Start playing these clips together and see which ones make each other sound better. Find three to four samples that fit together so well that they make you feel something: joy, sadness, anger, or the need to dance. Remove all the other samples from the project, but save them in a folder so you can access them at another time.
3. Now loop these samples linearly for at least 3 and a half minutes. This is the foundation of the house, and everything else that is being added to it should make this feeling you felt stronger.
4. Next, we will use the subtractive arrangement method to create a song structure. Start deleting sections of the sample you made so that you will have a clearer layout for where the song sections are. For example, maybe only bass and chords are in the intro, the verse might be just drums and bass, and all of them play together in the chorus. Try to make all of the sections also slightly different from each other to add variation.
5. Create drops and transitions by adding a couple of bars of quietness before and after sections. For example, deleting one to two bars off the drums and bass before the chorus can emphasise the song's dynamics.

**Walls of the house:**

6. This means you can listen to your creation and see what it needs. Maybe choruses need a higher arpeggiated synth to add hype or the synthesiser

sound you already added needs some stacking? Or perhaps you wish to add vocals? Make sure that the feeling you felt at the project's beginning is still there, and it will not get lost by adding too many different sounds and instruments. Remember, sometimes less is more.

**Decoration and character:**

7. This means you can start to apply effects, details, and automation. Maybe you wish to create tension with lifts, swells, and transitions? Or perhaps the vocals need layering or stacking, as well as other creative details? Give yourself time to play around with this part of the workflow and focus entirely on crafting all the elements you have brought into the session.

## Workflow 5: Collaboration Perspective

Collaborating with others can benefit our creativity, learning, and confidence. But so that we can feel like everyone's voices are heard equally and the creative input into the project is fair, you can use pre-agreed workflows as an aid. This minimises arguments and helps your team achieve an outcome everyone can feel happy and proud of.

The following workflow example is designed for a collaboration project where the creators are in the same room together and the outcome is not pre-defined.

**Yes. And... workflow:**

1. One person starts with a musical idea. This might be a chord progression, bass line, or just one line of lyrics.
2. The second person accepts this idea and takes it forward with their inspiration. This means that for a chord progression, they might add additional chord variation, accompany the bass line with a funky beat, or think of a melody sequence for the lyric given.
3. Everyone in the collaboration will take turns adding, changing, or removing something from the song, and it will be ready either when the time runs out or when the group decides it together.

**Other suggested workflow approaches you can take as a group:**

- **Distant Workflow:** If you collaborate with someone over the internet, you might adapt the workflow example mentioned above to an online-based scenario. You might want to work in the same DAW using

a session that can be shared or work using stems. This means that each time something new is added, the project will be exported to separate stems and sent to the collaborator.
- **Individual Brainstorming:** One way to work with others efficiently is to create a pool of sounds, samples and ideas pre-session. Then when meeting with the collaborators, a song can be formed from this material. If a further structure is needed to make the project easier to manage, one of the previous workflows from this chapter can be used as a guide for the group. For example, using the Genre-specific workflow to structure a song from the sounds can add needed direction to the teamwork.
- **Pre-defined Goal and Solving a Puzzle:** In some collaborations, it can be a good idea to have an agreed goal that defines the outcome. The starting point for the project is to figure out how to get to the outcome as a team. For example, if the goal is to make a pop song that sounds like chart music, the first things to figure out is what the style is constructed from and what individual strengths the group members have to accomplish these musical structures. Solving a puzzle like this can be inspiring and educational while being a great team-building exercise.

The next chapter will move from workflows to interviews with industry professionals. They will talk about their journeys into the music industry, insecurities and obstacles they might have felt, and how they overcame or are working towards overcoming them.

## Additional Resources

- 12 Days of Creativity Challenge: www.lnamusic.com/programs.
- LNA Does Audio Stuff. 2022. "How to make house track in 10 steps". YouTube. https://youtu.be/EDiZrTccfyg.
- LNA Does Audio Stuff. 2022. "How to make techno track in 10 steps". YouTube. https://youtu.be/8om2uak_FKU.
- LNA Does Audio Stuff. 2022. "5 ways to overcome creative blocks in music production". YouTube. https://youtu.be/1LxF7OzkOuA.
- LNA Does Audio Stuff. Download workflow templates from: www.lnamusic.com/shop.

# Industry Professionals Share 5

## Keywords

- gatekeeping
- gear

## Introduction

Thus far, this book has discussed and analysed the most common insecurities we experience in the audio and music industries. As much as psychological and sociological concepts can help us conceptualise and understand how our minds work, it is sometimes even more beneficial to hear about the things we struggle with from the people we admire and look up to. Even the most successful industry hero has felt the same fear, jealousy, or failure as you might have experienced. Reading and listening to others' side of the story can help us understand that we are all just humans, feeling all the same insecurities on our unique journeys. Everyone being interviewed in this chapter is brave to tell their personal stories, but they all have done it with intentions to give others hope, strength, and inspiration on their paths.

This chapter features interviews from ten industry professionals working in the music and audio industries as artists, music producers, composers, performers, engineers, social media content creators, and educators. The people chosen to be interviewed in this book come from different backgrounds and perspectives of the industry, giving as diverse a point of view as possible at the time of writing this book. But as the industry is big with so many amazing stories to be heard, it is good to point out that due to limitations, in these interviews, it was impossible to represent all

DOI: 10.4324/9781003194484-20

of the voices that should be heard. The topics of this book are often very personal, and that is why anyone talking about their insecurities publicly is brave, but also putting them in a vulnerable position. For example, people from minority groups may feel judged or worry their careers are at risk by opening up about their experiences. This is why, at the end of this chapter, you will find further resources featuring more interviews on similar topics.

In addition to this book, I wish to continue giving space for more discussions and interviews with artists about insecurities, confidence, and creativity through a podcast and a YouTube channel. The links to these can be found at the end of this chapter.

## Laura Escudé

She/Her
Artist, Music Producer, Performer, and Live Show Programmer/Designer

When I moved to Los Angeles in 2004, I had been out of school for a few years, and I knew that I wanted to come here and be a musician and perform. Performing was the scariest thing that I could ever imagine however I did one live show on my own in 2003. That was my first live solo performance and I remember feeling terrified to play in front of people. I would shake, and I was just so nervous. It was uncomfortable, but I knew that I needed to do it. When I moved to LA, I started getting opportunities to play and I tackled that fear head-on by playing show after show.

In early 2005 I got a job doing tech support at M-Audio, my first job in LA. One of the men immediately said that they just hired me because they wanted a woman in the room, which discounted the fact that I knew anything. And that stung and brought my confidence down a bit. But there were many supportive men at that time as well and they helped me to level up my skills because I didn't consider myself a very technical person before that. Then I made it my mission to become super technical because I felt I had to prove that I deserved to be there. I feared that unless I lead with confidence right away and proved myself, I would be treated like I did not know what I was talking about. So I immersed myself in learning everything. It was exhausting and so rewarding at the same time. After this experience I decided to make it my mission to help as many women break into the world of technology.

When I got into the touring world, it was also like that. I worked at Ableton in 2007–2008 and my first job on a show was in 2009 as a playback engineer. As I was one of the only Ableton Certified Trainers at the time, I had been doing many master classes and presentations – but I did not know what playback engineering was at the time. Playback engineering is a complete other way of using the software, very different from producing. So I went into this new job not knowing much about it yet. And I remember feeling humiliated because they brought in another guy very quickly and I felt they didn't give me a chance to learn or get trained. I had a contract, so I didn't get fired, but they had me just sit there, and I remember how unsupported I felt. It was heartbreaking because it was the biggest break of my life, and this was so huge. If someone had just shown me a few things, I know I would have gotten it very fast. It knocked my confidence down a lot because I knew Ableton Live very well, but this was an entire other skillset. I did not have the experience they needed, however I felt they could have dealt with it better. I vowed to be as open as possible with my knowledge moving forward and never gatekeep information. This led to me starting many different educational series geared towards helping folks become proficient in music playback.

At the same time, this experience lit a fire in me to learn how to be a playback engineer, and I started getting calls to do other shows. In 2009 I got connected with Kanye West's engineer and in 2011 I started working at arena shows for 20,000 to 30,000 people. I was terrified and remember being so anxious that I would screw up.

During these times, I was not aware of the anxiety and constantly living in fight or flight mode. When I did these big shows the way I released the stress was partying and drinking to relax and let it go. And now, I am not interested in any of that stuff. Now I want to go home and meditate and take a hot bath! So I reach for healthier ways to cope with the stress now, and I made a conscious decision to quit touring so much about six years ago when I had a big health crisis. I was in the hospital because I ran my body down, and after that, I realised that touring that much was not healthy for me anymore.

I also had versions of disordered eating for 20 years because I tried to make myself small, seeing images on TV and in magazines that I thought I needed to look like. I felt like maybe people would like me more if I was a certain size, or I would be more loved. Once I gave all of that up and let my body be the way it is naturally, I became so much happier. I could not imagine going back to that restriction level and focusing on how I looked all the time because it was exhausting.

In the past couple of years my journey has been about finding a healthy balance and I have written several articles online about my struggles and how I transitioned through them. I want to help people, especially women, by sharing my stories. And my music is now about sharing my expression, self-love processes and affirmations. Ultimately, I create music that helps nourish me, and it is my medicine. And If people are inspired by it and feel like they can express themselves or heal through listening to my music, that is so wonderful.

So this is what I have been experimenting with for the past couple of years – exploring a new understanding of myself and falling in love with who I truly am, not who I thought people wanted me to be. We are all reaching a new level of self-awareness as we get older and are on our healing journey. For so long, I was not ready for the journey, and I was still stuck in this idea of being perfect. And I was so afraid to not be perfect. But now, embracing my imperfections has allowed me to become more creative, as I do not have all that brain space taken up by trying to control everything. I celebrate how far I've come and love and accept myself with as much ease and grace as possible.

## Angel Lee

DJ and Producer

There have been moments where I have felt less valued and less respected, leading me to question my abilities. This generally stems from the DJ industry being male-led. There have been situations where I have been the only female within a DJ lineup and even the only female within particular working environments. Being treated differently from my male peers and being spoken to in a patronising way has often happened.

Moments like this have led me to doubt my abilities, making me question whether I am not treated the same because I do not deserve to be or because I am not as good a DJ as others. Situations like this have meant a knock on my confidence and self-belief.

I have learned to use instances like these to make me more determined to succeed. Any negative setbacks and comments make me want to work harder, learn more, and achieve more.

Keep going, and let any setbacks be the fuel to your success. Let any negative energy or situations help you become more determined. And most importantly, believe in yourself, and be your biggest cheerleader!

## Tim Linetsky

He/Him
Known also as Underbelly and You Suck at Producing (YouTube)
Producer, DJ, Pianist and YouTuber

I got to tour with this flute guy who went viral in the UK due to a meme. So we did a huge, five-week tour with 20 cities in the UK. And that was coming from no previous touring experience, and it was a dream come true. And then after that, I came back home, and I was so jazzed up, feeling like I was the king of the world. I felt like I do not need school and did not need to go back there, as it was a waste of time. I was just going to pursue to be an artist with 100%. And that is when I moved to LA with my friend.

But being in LA, not knowing anyone, not having a job or anything to do, and just sitting at home, trying to make music got me depressed. And nobody was hitting me back. None of the artists, agents, or managers gave a shit. So after a few months, I just could not take it anymore, and I moved back home and then went back to school. Thank God I did because, a week later, everybody in the house I was staying in got evicted anyway. So there would have been no future there. But that was certainly a humbling experience, especially after such a high before that.

So as far as believing that you need to be in a certain place to make things happen, I think it has never been more true that location does not matter. Especially for something like music, as you can do it all online and anywhere you are. For example, electronic music is all in the box, and you do not need anybody else, so all the networking, all the opportunities, can be done online.

But still, I think to get your foot in the door, you already have to have something pretty strong going for you. Because otherwise, nobody is not going to bother to have anything to do with you. So if you already have some career going online, I can understand why you want to move somewhere like LA. Because if you plug your brand into the LA music industry machine, Berlin or New York machine, there is an opportunity for growth there. But starting from the beginning in one of those places, you are just a small fish in an extremely large pond. Probably better to be a small fish in a smaller pond. You know, in your hometown.

I am not very deep in the industry and just starting to get label releases. So, the nice thing about my YouTube channel and the tutorials is that I can make a good living off music and make music without having to deal with all that industry bullshit. So I have been sort of industry adjacent for most of the time. But I think since LA, I have done well for myself, and the numbers have gone up, and more people know my name. So if I were to do the LA thing again, I think it would go better.

But I do not know what the music industry even means anymore. Certainly, the tippy top of the industry is still intact, the Taylor Swift and the Drake's of the world. And they probably still go to the Capitol Records offices sometimes. But other than that, there are so many different avenues now of gaining a following and putting yourself out there. So that level of gatekeeping has been erased, more or less. I do not think there is a traditional way to have an artist career anymore. Or at least there does not have to be. That being said, as I mentioned, I have been reaching out to labels. And it seems your name needs to be associated with certain other names and labels to be recognised wider and get booked for shows. But the beauty is that if you do not want to be in it, you do not have to. I could continue doing YouTube and putting out my stuff and not even bother with any of these people.

## Kelly Buckley

She/Her
Electronic Music Composer, Sound Designer, Producer

I was always into music since I was a really little girl. I started songwriting really, very young. I had guitar lessons from age about six to 16. That was, I suppose, a hobby. So as I got to sort of school leaving age, you are starting to think about what way am I going? I did love music, but back then, I do not think my working-class parents really understood how huge the music industry is. And that, I suppose, put me off the idea. And also, I am not blaming them. I mean, they just wanted the best for me.

And so, I went into journalism as my main job, and music became a hobby. I have been a journalist for around 28 years now. And as time went on, I was in bands, wrote songs ... was never good at playing instruments live, I never kept up the practise. So I always felt I have got to leave the instrument-playing and production to someone else for them to bring my songs to life. I brought the vocal melody in the session or the top-line, and they would produce it.

And as time went on, I got interested in soundtracks and started to think: how does that work? And of course, once you start going down that rabbit hole, there is a lot to learn about royalties, what you need to deliver and how it all works. So what started to happen is that I was trying to get everybody else in the band to register with PRS and do all this stuff with me. And it was quite hard to drag people along. And so somewhere along the line, I just thought, you know what, I really need to do this on my own.

I was interested in electronic music and was in a few electronic bands. I had dabbled with bits and bobs over the years, but with only basic stuff like GarageBand. I used to think I was not technical, leaving that to the boys. But then, becoming more interested in electronic music and watching these women on YouTube and Instagram, with their entire band in these boxes with buttons, I realised there is another wing to this story.

But then there was this band I was in called F\GUREHEAD, and this gig opportunity came up, which was quite well paid. And a boy that used to launch the beats live in the band ended up not being available for the gig, in fact a number of gigs and the rehearsals. So I just went and bought a Launchpad, and thought I have got to figure this out. And I was just using sort of royalty-free beats and loops so that we – myself and the other girl in the band – could get through this gig. That was the moment I realised: I can do this.

We carried on for a little while, but her priorities changed, and I had to go to a hospital for a major operation. So I had a lot of time off work; that is when I thought I needed to learn how to do this properly now. I joined the MPW (Music Production for Women) and did their courses. From that point, I started learning everything and doing as many courses as I could afford. The gear began slowly altering my live set so I could do everything on my own and create my own beats and loops out of found sounds.

And then COVID happened. I was put on furlough, but I knew I was going to get made redundant, so I took it. And so obviously, you get a little bit of money, even though it was not much, but it was enough to think I have to take this time. So I registered as self-employed and started my business as K-A-B productions. And just went through at it hard, building a licensing catalogue and trying to kind of put feelers out to let people know I could create music and sound design.

I am lucky I live in an area with a good art scene. And so there are a lot of people doing a lot of things like making films and what have you. There are a lot of artists, really good local galleries and art organisations and so people who I know just started to notice what I was doing music and sound-art and sound design and asked if I would be interested in doing things. I think the first thing I did was a soundtrack for an artist who had this exhibition in London. So then, one thing led to another, and I had quite a few good commissions just from that, even though I was still learning … am still learning.

I have got sons, and they are at an age now that they are not fully dependent on me anymore. When they were little, they could not do anything for themselves, but now they are 16 and 17. So, they are old enough for me to go; I am doing something for myself for a while. Sometimes, instead of going out on a Friday night, I will do music production. Because if you want to do something, you make time for it. You will put your social life off where you need to. Because when it is something you are obsessed with, you will need to make time for it.

The only thing where my age ever comes into it is when sometimes performing live. I have to sort of wrestle with myself sometimes as the little demons come through and say, "you probably look like some silly old woman now". But then I had to tell myself that I should do it because, why not? The world is full of these older guys doing this, pop and rock bands or even in the electronic music scene. It is normal to have guys in their 50s or even older up there still doing it. And so I think, why should I not?

So if you want to do music production and electronic music, you will find a way to do it because there is always a way, even if it is slow, even if you only have two hours a week that you can spare. Do it and put those two hours in because when you look back, you will feel proud that you did this, rather than just thinking about it and not ever doing anything about it. And make the most of it, life is short, and grab every opportunity. Or try and create opportunities and just bloody do not take yourself so seriously! Because nobody cares.

## Milton Mermikides

He/Him
Composer, Music Producer, Guitarist and Academic

I was not very good at school at all and I began to feel like I had no relation to education that made sense to me. I was super interested in everything, but at the same time, I did not fit in. And I was told I was not good at music when I was young because of some tests around it. And so I did not study music formally till much later in my life because I believed those measures, which was a mistake.

I studied at Berklee College of Music and thought I was able to do something. But then I arrived at this school, which is 60% international students. And you have New York saxophonists, Brazilian percussionists and incredible Japanese pianists trained in jazz. That talent and rich diversity were just so overwhelming.

And one of the biggest things was playing in time and whether people swing. I remember struggling with it so much I ended up crying in practice rooms. Criticisms cut deep, and I practised with my metronome with the lights turned off, and just this red metronome light flashing. And that tough lesson has now turned into a wonderfully rewarding lifelong obsession and academic pursuit into what makes things groove. So I think there can be insecurity events which can be catalysts to beautiful change.

I think that there are certain aspects of learning that literally hurt a little bit, and learning always involves some discomfort. And it is not a good thing to avoid that; we would never learn anything if we did not want to strive. But that does not have to involve deep anxiety. It can be a restless curiosity. When you have enough knockbacks, you are forced to do certain things so often that the anxiety becomes counterproductive and redundant because no one cares that much. So sometimes exposure to anxiety is the only cure to redressing it.

I mean, one thing to understand is that with all sorts of literacies, you can be both limited by and open to them. I see anxiety around people who do not understand technology, and I see it in classical musicians unable to improvise. And I see it around pop musicians not being able to read or not knowing the theory. And personally, my insecurity is around my singing, in which I am a good enough musician to know that I'm not a good singer. But as much as it would benefit anyone to push in "problem areas", you must also recognise what you can do. And I think the one thing that set me free is that I am a mixture of the literacies that I am interested in.

So when I worked at the Royal Academy of Music, I used to be called, whenever there was an electronic piece where you had to follow the score and make something happen at a certain point. I got it because I could follow a score and build a system to do it. Am I the best system builder in the world? Definitely not. Am I the best score follower in the world? Of course not. But the fact that I can do both at a certain level. And this is what I mean. We are a mix, a mess, a mosaic of literacies.

Similarly, what is eye-opening is just to see how many different ways people approach the same challenge. Then you start realising how much possible music there is to be made in the world, and it is infinite. I started thinking that when you make music, you think it is not original, and you might question if it is worth being there. But there is an abundance of music to be made and thoughts to be had, and it is never-ending. And that was a slow but huge, emergent realisation came to me when I understood there are no authorities, and that curiosity is always rewarded.

Overall, you should always make music to express yourself, not impress others. Of course, it matters if someone you admire likes it, but I do not think that strategy has ever worked for me. I used to care about how good I was, but I do not care about that anymore. I am too busy enjoying doing stuff. Again, life is too short to focus on impressing others.

I like to think of creating music as discovering, collecting, and organising seashells when you are on the beach. You might think, "aren't these beautiful?". And that is what I like to do with music; just present things that I find beautiful. And that is it. So if you honestly think something you have is beautiful, then share it.

## Ski Oakenfull

He/Him
Head of Education & Curriculum at Point Blank Music School

Ableton Certified Trainer, Artist, Keyboard Player, Producer, Remixer and Composer

I started playing the piano when I was six years old. My parents never said I should take up an instrument or forced me to practise. I started taking the piano grades, and I remember getting petrified on the day of the actual exam. I would experience cold sweats, and get so nervous.

It was at that point that I first became aware of the psychology around performing. When you are practising a piece in the lead up to an exam, you are just doing it without thinking, and it is a very relaxed mode. But then, as soon as you are put on the spot, you start to question your memory and ability to read music, and you can feel like you are just not doing it right or performing the piece to your best ability. When I was around 15 years old, I was in a band with my schoolmates, and we used to put on concerts for our friends. I used to have that similar feeling of just being so nervous, and it would physically affect me, giving me bad headaches.

I suppose it was because of fear of failure or being made to look stupid doing something in front of other people. I think it is much easier when you are in a band, when you are not front and centre, having people around you to cover up or mask any mistakes you might make. But when you are doing a solo piano concert, it is naturally very nerve-wracking. However, I think that this was very formative for me, and I am glad I had to go through that situation, because it was like breaking a barrier and pushing myself to perform.

Over the years I have found that there have been many situations where I had to push myself and throw myself into things. On some rare occasions I might have held back, but 99 times out of 100, having a cavalier attitude pays off and I haven't failed (whatever "failed" means!). Even if what I do is not perfect, it is not the end of the world, and the more I try, the more it becomes a real confidence booster, proving to myself that I didn't fail. Taking a risk makes me better equipped to deal with the next situation.

As much as I find tests and grades challenging, they create deadlines, structures, and boundaries for you, which can act as a useful steppingstone. When you start out as an artist, producer, or musician, it can feel like you are looking up at the top of a mountain and you'll never reach the peak. You might wonder how you will ever be as successful or technically as good as other people, which can demoralise you. So, I think if you set yourself goals in stages, and prove you can achieve each one, it will give you confidence to progress and get better.

I work well with deadlines which provide a limited amount of time to do some work, as it pushes me in a positive way. It's about setting something attainable in bite-sized chunks. From my experience of designing degree programs, it is a process of scaffolding the curriculum, where a student can't progress to the next level until they have proved they have met the learning outcomes for that particular level. For example, in sound design, you can't progress to "Reaktor Primary" you have learned the fundamentals of synthesis and you understand how to use "Reaktor Blocks". My kids have sometimes said to me that they don't understand why they have to do exams or why certain GCSE subjects are important. They are certainly learning some subjects that they will never touch again for the rest of their lives, so I can understand why they might ask "what is the point?". But overall, I think tests are a positive thing for learning.

If you fail tests, and they demoralise you to the point where you don't want to carry on, then I think they are obviously a negative thing. I have failed exams and it is not a great experience! But I believe in some cases, the issue might be around the assessment process. For example, to assess a violin performance there are many criteria from the tone, to finger technique and rhythm. That is why it is important when designing that type of assessment that you have a wide variety of criteria. Even if the performer might be lacking in one area, they may be good in another, so the assessment will balance out, and they will still get through it.

When it comes to studying music and production, I have seen some students who put the emphasis on acquiring as much detailed knowledge as possible and then try to show it all off in their tracks. This often means that the final result suffers from being unfocused and unmusical. From an academic perspective, the assessment of the work can be challenging because students are not only being graded on how "good" the piece of music is, but the technical aspects as well.

At the start of each term, I give an induction speech for new students, where I encourage them to work outside of their comfort zone. Often people might come to Point Blank having previously only worked in one or two different genres, so I talk about how important it is to explore other musical styles, to study and analyse different music. The focus then moves away from only needing detailed technical knowledge for the sake of it, and looking at more general production techniques. I'm not saying there's anything wrong with going deep into technical knowledge, but it's good to keep a more holistic, musical perspective. Maybe this is where having a good structure to the learning, with modules, deadlines, and being part of a program, is a good thing. It helps build confidence and focus, and avoids students going down a rabbit hole!

I think it's important to accept that you will never know everything, and that you are always learning. Even if you do have a detailed knowledge of certain subject areas, things always change as music and technology evolves with new research and innovation. People have said to me, "I cannot believe how much you know about music theory". That is their perspective because maybe they have not studied it before, but from my perspective I feel like I am just at the beginning because I know that I understand the basics, and I know I can always go deeper.

Over the years I have had to deal with trolls commenting on my YouTube tutorial videos, saying that I have made mistakes with my analysis or explanation, which forces me to go back and check the video. They might say something like, "that is not an augmented seventh chord, or that's a different mode, and you're completely wrong!". But after analysing it again and doing a bit more research, I may end up agreeing with them. I actually quite enjoy this process because it makes me think deeper. You need to put yourself on the line sometimes and learn how to deal with criticism.

In my role at Point Blank, I currently oversee more than four degree programmes, so it's got to the point where it's impossible to know every fine detail of each individual module. There are normally about 18 modules on a degree, so there's no way I can be an expert in every subject area (especially music business-related ones!). I think I have always had a slightly blinkered attitude towards learning. I will know enough for me to do what I want to do, and I don't particularly have a desire to learn something just for the sake of it. I think everything should be for a reason. However, I do see a massive benefit of exposing yourself to new techniques to get inspired and open opportunities, as this is about working outside of your comfort zone. I suppose I need to practise what I preach more often!

My ultimate way of spending time is to indulge myself in making music and enjoying the creative process. For me it is all about having a good workflow and being in an environment where I have things on hand so I can write and produce quickly. Feeling like I need to know every intricacy of programming in Max for Live isn't necessarily going to help me write better music, because I don't think it will push me to create. I suppose how much you want to know is what drives you.

## Rachel K Collier

Music Producer, Ableton Certified Trainer, Content Creator and Performer

Most of the darker moments in my career have all been music business and industry-related. Obviously, there have been a handful of times I felt insecure with my music, like in my recent production ABXY, where I was banging my head against the computer. However, that is a puzzle that I can work on and solve. Whereas the mystery of the industry, whether we are doing the right thing and questioning if this is the right career choice, is not so easy to solve. This will always just be trial and error. So when we talk about insecurities, it is more about how the industry would view me rather than what they think about my music.

The other day, I was at an event on a panel in London, which was amazing. There were some incredible people, and I was sitting there thinking, why am I here? Why am I on this panel? But when I talked to the guys who booked it, they said to me; "We booked you because you did not put your music out in the 80s". Then I understood that they brought musicians and producers from different eras. There have been many times when I felt like I didn't deserve to be on particular panels especially more during industry and networking events but I remind myself that there is a reason I am here and I haven't just been pulled off the street.

Obviously you have to hustle, and sometimes and you have to push for things. If I have to ask someone for something too many times, in the end I get this feeling: "you know what, screw them", because it's not always easy doing that. I kinda of say to myself If people want to work with me, they can come to me. I will not put myself in a situation where I am against 50 other writers, and maybe I will get picked.

It is awful how people are expected to go through that. I learned about it early in my career while doing a lot of top-line work, it was brutal. You have to be writing against 50 other writers and hope your songs get picked and make the cut. I was doing that for two or three years, I did have some success with it. I had a top 20 hit in the UK in 2013 called "Boom Boom Heartbeat" and also wrote "Only For You" with Mat Zo which was a Grammy-nominated record. But I still could not handle the mental side of it. I remember my music lawyer at the time saying to me: "Rachel, you need to write 100 songs, and for every 100 songs, one might make the cut".

I would go to certain studios, and the egos in those rooms would make me feel so inadequate. Even if I had just done something cool myself like travelling to Berlin to write the music for the Ableton 10 campaign, I would still come home and feel like absolute crap. None of it made me feel good about being a writer, a top-liner, or an artist. That is when I realised that I need to remove myself from these situations that make me feel useless.

Whenever I had a little bit of success, I would be saying to myself how I did not actually do any of the best parts of the song. Or how I did not write the best lyrics in that track. I was torturing myself. So then I went through this process of finding balance, and I got the book called *Power of Now*, and I swear by that book. It helped me with this overthinking mentality and the need to believe in myself. And actually, when these dark thoughts come into my mind, I actively practised saying, "No, piss off, I am here because I have the right to be here, and I am good at what I do".

Adjacent to trying to find confidence in my work and myself as an artist I started to produce my own music. It was the right decision for me because I got so much enjoyment from sitting alone in the studio and playing with plugins and synthesisers. Making the actual music has always been the thing that has kept me going throughout the toughest times in the industry.

Whenever it has been a dark time for me, it is always business related. It is never actually about music. When I think about those top-line sessions some of them were really painful. My own production sessions, which, of course, have sometimes been painful but for different reasons – problems that I can solve myself. To be honest, Ben, my manager, was brilliant, not just as a manager, but as a mental coach. He would tell me, "You are confusing the business with your passion, you are confusing numbers and figures that you cannot control with your love of making music". So I have tried to change my mindset over the years and realise that success is doing what you love every day.

Being able to sit here and have this time to make music, that is being successful. The business can never really satisfy you the same way as making music does. This might sound cheesy, but that is kind of it. I am trying to make sure with all my work that music is the main focus. As any current music independent artist knows, we have to spend a lot of time working on content, otherwise who the hell is going to hear the music in the end anyway?

My song, "You can pretend", has this attitude. If you give in to those insecurities, they will take you down. The book, *Power of Now*, has helped me do this. To become aware of my emotions and thoughts. Now I can find ways to back myself up, and stay true to myself. It might have been a slower journey, but at the end of the day I know I have not just followed a trend.

I know that there are ways that I could play into the trends to try and get more subscribers and win the algorithm gods, but I am not going to do that. I think whenever you try to do something for the wrong reasons, it will always backfire.

Just be yourself. Do not live for trends or anything that you think will just get you famous quick. As much as fame can quickly come along, it can disappear just as quick. If you craft and create something unique and special, you will build an authentic audience and real fans that will stick with you for much longer.

If I had just made straight-up EDM, it might have been more straightforward to promote my music – for example on places like Beatport or Tracksource. But I was sitting with my patrons the other day and they were literally saying that they love me because my music is not like anyone else's. This was what I needed to hear from time to time. It reassures me that I am doing something right.

So I would always say be true to yourself. If you want any chance to sustain your career, you need to love it, and be able to do it day in day out and devote your life to it. If a tiny part of you does not enjoy what you're making or creating, you will not be able to carry on. When you love it, you can do it forever.

## Taetro

Known on YouTube as Taetro.
Music producer, Electronic Musician, Controllerist, and Beat Maker.

One thing I recently started doing was I took most of the social media off of my phone. Instagram, the YouTube Studio app, TikTok, and Twitter do not live on my phone anymore. Because it was not just the idea of receiving negative comments but the constant access to feedback and continuous search for any validation. And having all that interaction with the work and direct access to all of that had detrimental effects. So, for example, if I posted a video, and it did well, my mood would be good for the rest of the day or the next couple of days. But if I posted a video that did poorly, it would affect my mood negatively. Asking myself, why am I in a bad mood right now? Realising I felt this way because I posted that video earlier. But since then, I have been trying to take steps to improve things for myself.

My mission is to either inspire people or enable them to be able to make their music. So my guiding Northstar is helping people, giving them some positive examples and giving them the tools I wish I had had when I was younger to learn music production. But if I can share my art with the world and help people teach them how to do the same things, that is great. I want to have a positive impact on the world in that respect.

When I put content out, that being the goal, it comes from a very positive place. So receiving negative feedback, I do not think I ever really feel insulted. Because I have learned that it is only random people on the internet, and I have demonstrated consistency with my community and shown expertise throughout doing this. I am pretty confident that what I am doing is both positive and educational.

And that is also where the positive feedback comes in. So once I had that internal talk with myself and had the confidence to believe what I was making was positive, educational, and good, the negativity had less effect.

But where the negative comments do affect me is when either just spreading misinformation or somebody is factually incorrect. But in an argumentative way, that can be frustrating. But with those comments, I have learned to click "Hide user from this channel", which is really empowering.

For me, subscriber numbers are not a goal, and I like to think that 1000 views per video are what I am happy with. But I will say it is a thing I must constantly remind myself of. So, where is my channel now if I do not get much more than 1000 views, something is probably wrong with the video in general. Because let us say I do have a video that does well and gets around 20,000 to 50,000 views. And then when you post the next video, which gets fewer views, it is easy to slip back into thinking, I guess I am not as good as that last video, and I need to make something else like that. So I am constantly back, reminding myself of my Northstar and telling myself to keep going.

I talk about doing YouTube as a producer, in contrast to the traditional route. And I think, in some ways, they are a bit similar. Because you work on songs, and hopefully, you get a placement, and somebody picks up some of your music. And in reality, that is almost as random as the YouTube algorithm. Except that it is based on people and their subjective taste, whereas the algorithm feels a bit more like a robot behind it. But either way, we are at somebody else's mercy, and there is some randomness to it. But with YouTube, if I stay consistent and build community rather than focus on getting music out to this wide audience, it can work out pretty well. And then also recentring my goals to my Northstar so I can constantly remind myself and not slip back into bad habits of putting myself down.

I have always been more interested in consuming stuff from people from other mediums, even though music is my medium. So for me, the video making is why this fell into this. I can make my music, but I can also capture the visuals of it. What is a good setting for this? How can I set up my space to shoot in multiple spots and make it look cool? And the process is what is fulfilling. If I were making more clinical videos, where I simply sat in front of a camera and taught it in a super procedural way, it probably would not be as fun. I put all the things I like together and enjoy the process. And I think that is why I have been able to stick with it so long.

At multiple points across this journey, I have taken moments to sit down, journal, and try to come up with answers to the question: What am I doing? What are my actual goals? What do I even want? Because when you sit down and write it down, it is out of your head, and it is on paper, you can either call yourself on your bullshit or have an affirmation for things you want. It also helps me remember, so I cannot forget, and I can always refer back to it.

When I started YouTube, I had no idea what I was doing. But I was not so clinical about it, thinking I had to post a certain way or on a specific day.

When it was a little more random, I had fewer followers and was just doing it because I thought it was exciting. And that is where my initial audience came in because they felt that energy. They felt something new and different. So I try to remember that if something is not working, do the opposite, go another direction, and reflect on the time when I just started.

I feel like I can speak confidently about these things, my Northstar, or my philosophy, but there are daily moments where I still feel slightly insecure. I still question myself, or I have to remind myself of the goals of my Northstar. So it is not like I have got it all figured out – far from it. It is constant daily reminders about what is right for me, battling internal negative thoughts and trying to push them down. But I have learned that over time, the more you do that and practise the positive habits, the more those positive mindsets take over and win the battle.

## Alice Yalcin Efe

She/Her
Electronic Music Producer and YouTuber
Founder of Mercurial Tones Electronic Music Academy

End of 2020 when my channel started to really grow. That is when I began to feel that expectation that now, when I make a release or any track, it has to be on top of the game. So I cannot just make a track that I enjoyed making but absolutely do it to perfection. This led to the biggest break in my life without music, not making tracks and putting them out. And I always go back and consider why I let that insecurity take hold of me. It is one of the negative effects of being a YouTuber because you are in front of people, and you teach production. So when you make your own track, there is no room for error.

But of course, that is completely not true. But that is the feeling that you have. It feels like you did something wrong. The expectations are high, and you have to make the top track and best-sounding mix each time. And that led to a year of not producing anything, which was a nightmare. But you can make a bad track, and it is okay. Even the biggest artists make bad tracks that are not that good. It's fine.

The industry is very judgemental, from the point of that, that you have to make everything perfect. And even if you do everything perfectly, people still find things buggy because maybe they are listening with bad headphones or listening from mobile phones or simply don't like the track. But the thing is, there will be tons of people who will relate to your music, and many will not be not enjoying it. That's just how it is with art. There are always lovers and haters. You cannot stop it.

The other part of social media is that hate comment that you receive. But you have no idea about that person. That person could be just seven years old, a frustrated schoolboy who is angry with the teacher and trying to vent it off. And your videos are the first ones the person is watching, and you have been the price. Or maybe another person who is living a completely different truth with the things in their life. And that is the problem that you never know how to react. If they were a person I knew, I would know how to respond because I know what my friend thinks, as we live in the same bubble. So we can understand each other a bit more. But I have no idea who is behind the hate comment, so it is tough to make sense of them.

I separate the comments into different parts. So if I see a person go straight after me, I do not care anymore. But let's say some person comes to my comment section and talks with bad intentions to other artists. I just ban that person from the channels because those types of things can escalate quickly. I do not do it for myself but for other people in my community, so we keep it in a safe space. So the platform does not become a tool to spread hate. I can handle it, but I do not know if that person receiving the hate will.

If I see a comment that is not possibly written with bad intentions but is still not good, I try to ignore it all the way. I also realised that if you ban or delete that type of person, they get frustrated, which can turn back to you negatively. But letting them be answered nicely helps them make sure the next time they might think about their action. So giving them that opportunity to learn creates a more constructive space.

As a musician and content creator, you get constant feedback. So I have to grow thick skin somehow, so I can live with it, because this life now is my full-time job, and I have to earn my income this way. I have to believe the pros and cons, and the cons come with a lot of insecurities because of the direct feedback. If I get something negative every time my new video is out, I get anxious and insecure. Did I teach it correctly there? Did I make a stupid mistake? And will I be judged afterwards? Every time a video goes up, you have this insecurity about the feedback from the people and the algorithm itself.

Whether it is TikTok, Instagram, those algorithm arrows, or view counts, even the best of us still will be affected by them. Because we have direct feedback from thousands of people in a second. So when I post the video, I see the views and how many people liked it. And on top of that, I will see how the YouTube algorithm liked it and whether it will be like it pushed forward for more people or not. And living with that constant feedback is very draining. Some people will be affected more, and others will be affected less.

It is also good to remember that every insecurity is key to improving your life. When I was about eight years old, it was my first time on the stage in this primary school. We had this big show where I was a singer and felt so insecure. Maybe around a thousand people were watching the event, and then I remember how I was shaking and thinking about how my singing voice is not good. But I did it, and that was a changing moment for me. I decided I would be making music for the rest of my life. If I never took that opportunity then and went after my insecurity, I would probably never be now in this moment, in this space. So I suggest people think about how their insecurities could be the key to their next improvement or the next big thing in their life.

## Additional Resources

I will continue these conversations on my podcast and YouTube channel. You can find me there: www.creative-confidence.com.

- Channel 4 News. 2017. "Stormzy interview (extended): Dealing with depression while making his new album". YouTube. https://youtu.be/nUMyMJayhbc?si=lw3SkMcGvXqq26ER.
- Giono, Y. T. 2021. "David Bowie's advice to young artists: Excerpts from the Michael Apted DVD 'Inspirations'". YouTube. https://youtu.be/JRtZc_Nmo5w?si=7Mx_1hcCFRVz0Egd.
- Huang, A. 2024. *Make Your Own Rules: Stories and Hard-Earned Advice from a Creator in the Digital Age.* S&S/Simon Element.
- Lamont, T. 2014. "Jamie T: Whatever happened to the Likely Lad?" *The Guardian.* www.theguardian.com/music/2014/sep/21/jamie-t-interview-carry-on-the-grudge.
- Marie, K. ed. 2022. *Conversations with Women in Music Production: The Interviews.* Backbeat.
- Netflix. 2023. *Lewis Capaldi: How I'm feeling now.* Documentary.
- Paramore. 2023. "Paramore – 'This is why'". Interview with Apple Music & Zane Lowe. YouTube. https://youtu.be/N4LxD8rf9Po?si=cAEvk6CT1L7zF9q9.
- Spanos, B. 2021. "Ms. Lauryn Hill speaks in depth about fame, racism, and 'Miseducation'". *Rolling Stone.* www.rollingstone.com/music/music-features/lauryn-hill-rare-interview-miseducation-500-greatest-albums-1109491/.
- Shukri, A. 2021. "A conversation with … Color of music collective, Alex Shukri". *Square One Magazine.* www.squareonemagazine.co.uk/interviews/a-conversation-with-color-of-music-collective.

# Acknowledgements

My husband Stephen, I cannot thank you enough for everything you have done to support and help to make this a reality. Thank you for listening to my endless creative confidence monologues (I mostly think aloud), the cups of tea and biscuits that kept me going and especially for translating my thoughts into text when my dyslexic, non-English brain could not do it. But most importantly, thank you for believing in me when I struggled to do it myself.

To my family, friends, and therapist, thank you for all the deep conversations and the long walks where hours were spent analysing our lives. Without you, it would have been impossible to reflect, the writing journey would have been lonely and topics left hollow.

Thanks to the whole Routledge team and especially my editor, Hannah, for publishing this book. Without you, I would not have considered writing this book and sharing my passion for creative confidence.

# Glossary

**Ableton Live**  Music production software (DAW) developed by Ableton.
**anxiety**  A feeling of worry, fear, and unease. Anxiety is a general term for this negative emotion, which we all experience at certain points in life. When anxiety becomes harder to control and starts affecting a person's everyday life, the condition is called Generalised Anxiety Disorder (GAD).
**audio software**  A computer interface that allows users to make, record, and create music. Also known as a DAW (Digital Audio Workstation).
**authenticity**  An authentic person is in harmony with their passions and values, not caring about external opinions.
**brain waves**  Electrical impulses in the brain.
**career fear**  Not feeling deserving of success in one's goals.
**compressor**  An audio effect tool that allows the user to control the dynamic range of signals.
**confidence**  The belief that a person can achieve their goals and plans, based on their abilities, skills, and self-worth.
**conscious mind**  The ability to think and talk about thoughts, feelings, wishes, and memories rationally while being aware of them.
**convergent thinking**  The process of analysing, organising, and using provided material. This linear thinking model is often used in problem-solving.
**creativity**  The use of imagination to create or invent something.
**DAW**  Acronym for Digital Audio Workstation; a music-making computer software.
**defence mechanisms**  Unconscious behaviours that protect a person from events that can create uncomfortable feelings, thoughts, and actions.

## Glossary 175

**divergent thinking**   Reaching for ideas without judgement or analytic capacity. This non-linear thinking model leads to the formation of new and original thoughts.

**elitism**   The belief that a society or group should be led by an elite group of people.

**emotional bypassing**   When a person does not allow themselves or others to fully process and feel negative feelings.

**envy**   A negative feeling of longing for something someone else has or for their luck or qualities.

**equaliser**   An audio effect tool that allows the user to control the frequency range of signals.

**failure**   Feeling a lack of success from not achieving a desired goal.

**fear**   A negative or unpleasant emotion caused by awareness of danger.

**feedback loop**   An effect that occurs when a person uses and interacts with a computer or interface. It consists of an action and effect, which can have a positive, balancing, or negative impact on the user.

**flow**   A feeling or state of being immersively focused and involved in an activity.

**gatekeeping**   A person or group limiting, controlling, criticising, or defining a section or sections of the music industry.

**gear**   A term commonly used for electronic music-making equipment, controllers, tools, and instruments.

**haptic interaction**   The physical experience of interacting with an interface, often through movement.

**identity**   A person's (or group's) beliefs, personality traits, expressions, and qualities that describe them.

**imposter syndrome**   Self-doubt that makes a person feel inadequate and fraudulent in what they do.

**insecurity**   The lack of confidence; uncertainty or anxiety about oneself.

**interface**   A device, program, or software that allows the user to use and interact with the computer or system.

**jealousy**   A feeling of unhappiness, anger, fear, concern, and insecurity about losing something a person feels ownership over.

**meditation**   A practice where a person aims to concentrate on specific points of focus and finds awareness in their mind and body. Different meditation techniques, such as Mindfulness, can be used to find clarity in everyday situations.

**MIDI controller**   A device to physically control MIDI instructions. MIDI controllers usually have keys, like a piano, square pads, or knobs, to control parameters in the software.

**MIDI**   Acronym for Music Instrument Digital Interface. It is a way for a person to send computer instructions on when and how to play virtual instruments in music-making software.
**modular synthesiser**   A synthesiser put together from separate modules with different functions.
**music theory**   The study of understanding the possibilities of music.
**musical genre**   A category that identifies a piece's musical style according to its characteristics.
**perfectionism**   A psychological concept for someone who has unrealistic demands on goals and believes they should be flawless. Perfectionists do not accept failure, experiencing worthlessness.
**plan**   A manuscript that lays out every part of your goals and aims into a clear and realistic path. Plans allow you to see your values, goals, and definition of success as a big picture, like a map.
**plugin**   Audio plugins are additional, often third-party, devices that can be installed on your music-making software. They can be audio effects or virtual instruments.
**procrastination**   Postponing or delaying something that should be done. This is an unnecessary voluntary action.
**self-belief**   The confidence in your skills, abilities, and judgement.
**self-esteem**   A person's belief in their own worth or abilities.
**self-regulation (also emotional regulation)**   Being able to handle your feelings, thoughts, and actions based on what's happening inside you and around you. Like having a mental toolkit to manage your emotions, thoughts, and actions.
**sense of self**   One's collection of beliefs about themselves.
**shame**   A feeling of humiliation or regret causing distress.
**sidechaining**   A music production technique where one signal is modulated with another signal using audio effect tools to control the effect.
**social identity threat**   When a person downplays their skills by believing they are incapable due to identifying themselves as part of a particular group. Also called VST's (Virtual Studio Technology).
**STEM identity**   Someone who sees themselves as technologically, scientifically, and mathematically inclined. They believe they have the capabilities to learn and practice all of these topics.
**STEM**   Acronym for science, technology, engineering, and mathematics.
**stereotype threat**   Similar to Social Identity Threat, it is when a person believes in stereotypes towards themselves and starts to conduct the behavioural patterns stated by these stereotypes.
**techy brain**   A concept often used to describe a technologically minded person.

**The Quartet of Creativity**   A method to demonstrate the effects of the world and sciences on our creative confidence. This is explained further in Chapter 1.

**time blocking**   A time management technique to help organise activities in the schedule, prioritise, and make space for things that might otherwise be neglected.

**unconscious mind**   The thoughts, feelings, wishes, and memories that are unavailable for us to examine. Whatever exists at this level of our mind often affects our conscious mind without us being directly aware of it.

**validation**   To validate someone's feelings is the act of listening and recognising how they feel so that the person experiences that their emotions are being heard and they matter.

**workflow**   A repeated activity pattern that happens naturally or is pre-planned. Workflows provide a way to organise and dissect information and resources, guide, teach, and be used as a problem-solving method.

# Index

Ableton Live 28, 30, 32, 155, 165; *see also* audio software; DAW
accessibility 26
anxiety 13, 29, 55, 92, 99, 106; *see also* fear; insecurity
authenticity 13, 60–62, 89, 111, 112

Belle, S. 61, 97, 107, 112, 124
Buckley, K. 48, 159

career fear 91, 94
Collier, R. K. 60, 165
confidence 12–14; audio spaces and gender 34–40; flow and confidence 16–21; insecurities 14–16; production tools 32–33
creativity 9–32; interface design and creativity 29–32; technology and creativity 26–29; *see also* flow; The Quartet of Creativity

DAW 6, 8, 28, 79, 145; *see also* audio Ableton Live; software
defense mechanisms 35, 43–44
divergent thinking 18, 141, 142; *see also* convergent thinking
Dobson, L. 34, 35

emotional bypassing 113–114, 121
envy 105–109
Escudé, L. 155

failure 121–126
fear 121–126; *see also* anxiety
feedback loop 21, 31
flow 16–21; *see also* creativity

gatekeeping 53–55, 67, 102
gear fetishism 76, 80
Greenwood, T. 13, 92

haptic interaction 7, 29–31
Helsinki Bus Stop Theory 67
human–computer interaction 30

identity 29, 36, 38, 47–50; social identity threat 47; stereotype threat 47–48; *see also* STEM identity
imposter syndrome 35–38, 48, 50; *see also* insecurity
insecurity 15–18, 43
interface design & creativity 29–32

jealousy 97, 105–110
Jepson, R. 12, 13

Lee, A. 157
LGBTQ+ 36
Linetsky, T. 157
LNA 71, 72
LNA Does Audio Stuff 80

Mark, T. 36
Mermikides, M. 132, 161
MIDI 27, 30
modular synthesizer 27, 28, 55
music theory 73–76, 144

Oakenfull, S. 94, 162

perfectionism 97–103, 121, 125
plan 127–130
procrastination 91–95, 141
productivity 92–95

The Quartet of Creativity 7–9; *see also* creativity

Rogers, S. 17, 19

self-belief 92; *see also* confidence

self-esteem 12, 56
self-regulation 115
shame 97–98, 112, 116
social identity threat 47; *see also* stereotype threat
STEM 11, 24
STEM identity 24, 28, 38, 47–49; *see also* identity
stereotype threat 47–48

Taetro 168
techy brain 47–51
time blocking 95, 140
Tricaud, M. 27
Tuck, J. 30

unconscious mind 81
user experience 29–32

validation 97, 111–121

workflow 18, 19, 80, 127–133

Yalcin Efe, A. 170

For Product Safety Concerns and Information please contact our EU representative GPSR@taylorandfrancis.com Taylor & Francis Verlag GmbH, Kaufingerstraße 24, 80331 München, Germany